SIGNIFICANT CHANGES TO THE

INTERNATIONAL FIRE CODE®

2015 EDITION

SIGNIFICANT CHANGES TO
THE INTERNATIONAL FIRE CODE®

2015 EDITION

International Code Council

ICC Staff :

Executive Vice President and Director of Business
Development: Mark A. Johnson

Senior Vice President, Business and Product
Development: Hamid Naderi

Vice President and Technical Director, Education
and Product Development: Doug Thornburg

Director, Project and Special Sales: Suzane Nunes

Senior Marketing Specialist: Dianna Hallmark

ISBN: 978-1-60983-567-5

Cover Design: Lisa Triska
Project Head: Steve Van Note
Publications Manager: Mary Lou Luif

COPYRIGHT © 2014

INTERNATIONAL
CODE COUNCIL®

Errata on various ICC publications may be available at www.iccsafe.org/errata.

First Printing: July 2014

PRINTED IN THE U.S.A.

Contents

Preface

The purpose of *Significant Changes to the International Fire Code®*, 2015 Edition, is to familiarize fire officials, building officials, plans examiners, fire inspectors, design professionals and others with many of the important changes in the *2015 International Fire Code* (IFC®). This publication is designed to assist code users in identifying the specific code changes that have occurred and, more important, in understanding the reasons behind the changes. It is also a valuable resource for jurisdictions in their code adoption process.

Only a portion of the total number of code changes to the IFC are discussed in this book. The changes selected were identified for a number of reasons, including their frequency of application, special significance or change in application. However, the importance of the changes not included is not to be diminished. Further information on all code changes can be found in the Code Changes Resource Collection, available from the International Code Council® (ICC®). This resource collection provides the published documentation for each successful code change contained in the 2015 IFC since the 2012 edition.

Significant Changes to the International Fire Code, 2015 Edition, is arranged to follow the general layout of the IFC, including code sections and section number format. The table of contents, in addition to providing guidance in the use of this publication, allows for a quick identification of those significant code changes that occur in the 2015 IFC.

Throughout the book, each change is accompanied by a photograph or an illustration to assist in and enhance the reader's understanding of the specific change. A summary and a discussion of the significance of the change are also provided. Each code change is identified by type, be it an addition, modification, clarification or deletion.

The code change itself is presented in a legislative format similar to the style utilized for code change proposals. Deleted code language is shown with a strikethrough, and new code text is indicated by underlining. As a result, the actual 2015 code language is provided as well as a comparison with the 2012 language, so the user can easily determine changes to the specific code text.

As with any code change text, *Significant Changes to the International Fire Code,* 2015 Edition, is best used as a companion to the 2015 IFC. Because only a limited discussion of each change is provided, the reader should reference the code itself in order to gain a more comprehensive understanding of the code change and its application.

The commentary and opinions set forth in this text are those of the authors and do not necessarily represent the official position of ICC. In addition, they may not represent the views of any enforcing agency because such agencies have the sole authority to render interpretations of the IFC. In many cases, the explanatory material is derived from the reasoning expressed by code change proponents.

Comments concerning this publication are encouraged and may be directed to ICC at significantchanges@iccsafe.org.

About the *International Fire Code*®

Fire code officials, fire inspectors, building officials, design professionals, contractors and others involved in the field of fire safety recognize the need for a modern, up-to-date fire code. The *International Fire Code*® (IFC), 2015 edition, is intended to meet these needs through model code regulations that safeguard the public health and safety in all communities, large and small. The IFC is kept up to date through ICC's open code development process. The provisions of the 2012 edition, along with those code changes approved through 2013, make up the 2015 edition.

One in a family of International Codes® published by ICC, the IFC is a model code that regulates minimum fire safety requirements for new and existing buildings, facilities, storage and processes. It addresses fire prevention, fire protection, life safety and safe storage and use of hazardous materials. The IFC provides a total approach of controlling hazards in all buildings and sites, regardless of the hazard being indoors or outdoors.

The IFC is a design document. For example, before a building is constructed, the site must be provided with an adequate water supply for fire-fighting operations and a means of building access for emergency responders in the event of a medical emergency, fire or natural or technological disaster. Depending on the building's occupancy and uses, the IFC regulates the various hazards that may be housed within the building, including refrigeration systems, application of flammable finishes, fueling of motor vehicles, high-piled combustible storage and the storage and use of hazardous materials. The IFC sets forth minimum requirements for these and other hazards and contains requirements for maintaining the life safety of building occupants, the protection of emergency responders, and to limit the damage to a building and its contents as the result of a fire, explosion or unauthorized hazardous material discharge and electrical systems. The IFC is available for adoption and use by jurisdictions internationally. Its use within a governmental jurisdiction is intended to be accomplished through adoption by reference, in accordance with proceedings establishing the jurisdiction's laws.

Acknowledgments

Significant Changes to the International Fire Code, 2015 Edition, is the result of a collaborative effort, and the authors are grateful for the assistance and contributions by the following talented staff of ICC Business and Product Development: Hamid Naderi, P.E., C.B.O., Senior Vice President; Doug Thornburg, A.I.A., C.B.O., Vice President, Director of Education and Certification; and Jay Woodward, A.I.A, Senior Staff Architect. The authors would also like to thank Tom Hammerberg, President/Executive Director of American Fire Alarm Association; Howard Hopper, Regulatory Services Program Coordinator with Underwriters Laboratories LLC; John Taecker, Senior Regulatory Engineer with Underwriters Laboratories LLC; and Bob Davidson, Davidson Code Concepts, LLC, for their insight and assistance with application of some of the technical requirements.

About the Authors

Fulton R. Cochran, CBO, CFCO

Deputy Fire Marshal – Retired

Henderson, Nevada

Fulton has been involved in the fire service for over 35 years. He started as a volunteer firefighter in his hometown of Manitou Springs, Colorado. Wanting a career in the fire service, he worked as a career firefighter and then moved into fire prevention. He was Fire Marshal in Breckenridge, Colorado, and an active member of the Fire Marshals Association of Colorado prior to moving to Southern Nevada. For the past 20 years, Fulton was with the City of Henderson, Nevada, where he was the Deputy Fire Marshal – Engineering. During his tenure, the City of Henderson was the fastest growing city in the country for over 10 years. During this explosive growth, Fulton managed the fire plan check team that reviewed the following:

- Master planned communities with over 10,000 homes
- Major infrastructure projects such as a 600-MGD water treatment plant
- Major commercial sites such as a regional mall and large strip centers
- Regional distribution centers greater than 750,000 square feet with high-piled storage
- Chemical plants with multiple hazards
- High-rise hotels and casinos
- High-rise hospital and medical office complexes.

Fulton has participated in code development starting with the legacy organization ICBO, and he attended his first code development hearing in 1985. Fulton has worked with the International Code Council (ICC) on numerous committees and councils. He was the International Association of Fire Chiefs (IAFC) representative on the Performance Code drafting committee. In 2007, he was appointed to the Fire Code Council, and in 2010 he was elected as Chairman of this Council. When ICC reformulated the Councils in 2011, creating the current Fire Service Membership Council (FSMC),

Fulton was again selected by the governing committee to be the Chairman of the Council, a position he continues to hold after being reelected most recently in Atlantic City during the 2013 Annual Business Meeting.

Fulton is currently a member of the ICC Board for International Professional Standards (BIPS), which oversees the ICC certification and testing programs. He represents the fire service on the Codes and Standards Council, which advises the ICC Board regarding the technical code committees and code development. Fulton was a member of the cdpACCESS steering committee charged with developing the framework of this program and recommending these actions to the ICC Board of Directors.

In 2013, Fulton was honored by ICC as the recipient of the Fire Service Person of the Year award.

During his tenure at Henderson, Fulton has been involved in numerous code adoptions with extensive local amendments. He is an active member of the Southern Nevada Fire Code Committee and is a two-term past President of the Southern Nevada Chapter of ICC. Fulton is also the former chair of the EduCode Committee of the Southern Nevada Chapter for code training and professional development.

Kevin H. Scott

President

KH Scott & Associates LLC

Kevin Scott is President of KH Scott & Associates LLC. Kevin has extensive experience in the development of fire safety, building safety and hazardous materials regulations. Kevin has actively worked for over 25 years in the development of fire code, building code and fire safety regulations at the local, state, national and international levels. Kevin previously worked as a Senior Regional Manager with the International Code Council, and before that, he was Deputy Chief for the Kern County Fire Department, California, where he worked for 30 years. He has developed and presented many seminars on a variety of technical subjects including means of egress, high-piled combustible storage, hazardous materials, and plan review and inspection practices.

Kevin was a member of the original IFC Drafting Committee that worked to create the first edition of the IFC. He served for seven years on the IFC Code Development Committee and was chairperson for the committee from 2001 to 2004. Kevin has actively participated in numerous technical committees to evaluate specific hazards and technologies, and create regulations specific to those hazards. Some of the more significant committees are

- High-piled Combustible Storage Committee
- Hydrogen Gas Ad Hoc Committee
- Task Group 400
- Technical Advisory Committee on Retail Storage of Group 'A' Plastic Commodities
- Underwriter's Laboratories Fire Council.

Kevin's constant work to improve fire and life safety has been recognized on many levels. His contributions have been acknowledged by various organizations when they presented him with the following awards:

- Mary Eriksen-Rattan Award in 2013—presented by the Southern California State Fire Prevention Officers' Association
- William Goss Award in 2009—presented by the California State Firefighters Association
- Fire Official of the Year Award in 2005—presented by the California Building Officials
- Robert W. Gain Award in 2003—presented by the International Fire Code Institute.

About the International Code Council®

The International Code Council is a member-focused association. It is dedicated to developing model codes and standards used in the design, build and compliance process to construct safe, sustainable, affordable and resilient structures. Most U.S. communities and many global markets choose the International Codes. ICC Evaluation Service (ICC-ES) is the industry leader in performing technical evaluations for code compliance fostering safe and sustainable design and construction.

ICC Headquarters: 500 New Jersey Avenue, NW, 6th Floor, Washington, DC 20001

Regional Offices: Birmingham, AL; Chicago, IL; Los Angeles, CA 1-888-422-7233 (ICC-SAFE) www.iccsafe.org

1

Administration

Chapters 1 and 2

- **Chapter 1** Scope and Administration
  No changes addressed
- **Chapter 2** Definitions No changes addressed

T he provisions of Chapter 1 set forth requirements for the adoption, application, enforcement, and administration of the *International Fire Code* (IFC). In addition to establishing the general scope and intent of the IFC, the chapter addresses construction, operational and maintenance inspection authority, the responsibilities and authority of the fire code official, and the adoption of referenced standards. Upon adoption by a governmental jurisdiction, Chapter 1 establishes requirements for the issuance of construction and operational permits, inspection of facilities and processes, collection of fees, issuance of stop-use orders and other legal notices, and formation of a board of appeals.

Chapter 2 provides definitions for terms used throughout the IFC and the *International Building Code* (IBC). Codes are technical documents, and every word and term can add to or change the meaning of a provision and its application. ■

PART 2

General Safety Provisions

Chapters 3 and 4

- **Chapter 3** General Requirements
- **Chapter 4** Emergency Planning and Preparedness

<u>**312.3**</u>
Vehicle Impact Protection

<u>**315.6, 605.12**</u>
Storage and Abandoned Wiring in Plenums

<u>**403**</u>
Emergency Preparedness Requirements

A basic requirement of the *International Fire Code* (IFC) is to prevent the ignition of materials inside and outside buildings. Controlling fuels and ignition sources limits the potential for fire. Chapter 3 contains requirements for combustible waste materials, control or elimination of ignition sources, open flames and recreational fires, and the use of smoking materials. Certain equipment can also be a source of ignition, and Chapter 3 addresses the proper operation of asphalt kettles and powered industrial trucks. In occupancies such as assembly uses of covered malls, controls are specified for certain hazardous materials or displays of vehicles. Chapter 3 also contains requirements for the protection from vehicle impact to prevent the release of compressed gas, flammable liquid or hazardous materials.

Chapter 4 requires that evacuation plans be prepared, that a hazardous materials communication program be established, and that employees be trained to identify fire hazards and safely evacuate building occupants. ■

CHANGE TYPE: Addition

CHANGE SUMMARY: This change authorizes the code official to approve barriers other than posts.

2015 CODE:

SECTION 312
VEHICLE IMPACT PROTECTION

312.1 General. Vehicle impact protection required by this code shall be provided by posts that comply with Section 312.2 or by other approved physical barriers that comply with Section 312.3.

312.2 Posts. Guard posts shall comply with all of the following requirements:

1. Constructed of steel not less than 4 inches (102 mm) in diameter and concrete filled.

2. Spaced not more than 4 feet (1219 mm) between posts on center.

3. Set not less than 3 feet (914 mm) deep in a concrete footing of not less than a 15-inch (381 mm) diameter.

4. Set with the top of the posts not less than 3 feet (914 mm) above ground.

5. Located not less than 3 feet (914 mm) from the protected object.

312.3 Other Barriers. ~~Physical barriers shall be a minimum of 36 inches (914 mm) in height and shall resist a force of 12,000 pounds (53 375 N) applied 36 inches (914 mm) above the adjacent ground surface.~~ Barriers other than posts specified in Section 312.2 that are designed to resist, deflect or visually deter vehicular impact commensurate with an anticipated impact scenario shall be permitted where approved.

CHANGE SIGNIFICANCE: The 2012 language in Section 312.3 is a carryover from the merger of the three legacy codes. This section was originally from the BOCA *National Fire Prevention Code*. The initial intent of the code was to provide prescriptive design criteria in Section 312.2 and performance-based criteria in Section 312.3, but it did not provide for a true performance-based alternative since it addressed a force, but not an impact velocity (i.e., a 12,000-pound vehicle traveling at 10 mph).

The change for the 2015 IFC provides a true performance-based option. The code official now has the ability to approve a barrier based on its ability to resist, deflect or visually deter vehicular impact.

312.3

Vehicle Impact Protection

Alternative vehicle impact protection

315.6, 605.12

Storage and Abandoned Wiring in Plenums

CHANGE TYPE: Addition

CHANGE SUMMARY: This change prohibits storage in air-handling plenums. Abandoned material and wiring cables must be removed from plenums.

2015 CODE: 315.6 Storage in Plenums. Storage shall not be permitted in plenums. Abandoned material in plenums shall be deemed to be storage and shall be removed. Where located in plenums, the accessible portion of abandoned cables in plenums that are not identified for future use with a tag shall be deemed storage and shall be removed.

605.12 Abandoned Wiring in Plenums. Accessible portions of abandoned cables in air-handling plenums shall be removed. Cables that are unused and have not been tagged for future use shall be considered abandoned.

SECTION 202
GENERAL DEFINITIONS

Plenum. An enclosed portion of the building structure, other than an occupiable space being conditioned, that is designed to allow air movement, and thereby serve as part of an air distribution system.

CHANGE SIGNIFICANCE: Common sense dictates that plenums are not suitable for storage. However, until this section of code was added, there was no specific prohibition against storage in plenums in the IFC. Storage

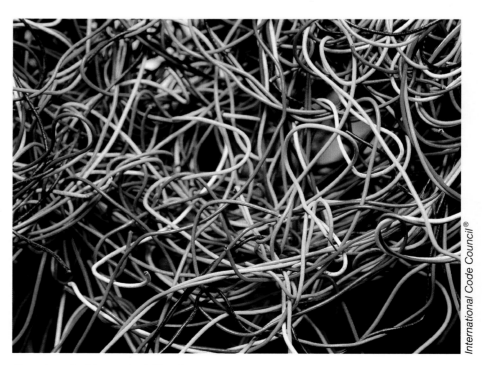

Abandoned wiring material in plenum

International Code Council®

in plenums adds fuel load to an area that may not be sprinklered, can cause a ceiling to collapse under a fire condition and adds additional risk to firefighters as this storage represents an unknown threat.

This new section is intended to introduce a concept that has been in the NFPA 70 *National Electrical Code*® (as well as in NFPA 90A, Standard for the Installation of Air-Conditioning and Ventilating Systems) for a long time: plenums are intended for a specific use as stated in the definition. That is, they are to be a part of the air distribution system so as to allow air movement. Plenums are also used (legitimately) for stringing communications and data cables as well as pipes and sprinkler pipes and other similar products. However, it is a common construction practice not to make the effort to remove wiring and cables when they become obsolete.

The proposal requires only the "accessible portions" of abandoned cables be removed, because there is no intent to cause potential damage to the building or facility by attempting to remove cables or circuits that are strung through walls, floors or other building elements. Conversely, unused cables that are tagged for future use are allowed to remain.

403

Emergency Preparedness Requirements

CHANGE TYPE: Modification

CHANGE SUMMARY: This section has been extensively revised and the content updated for consistency. To assist the fire code official, many portions of this chapter have been relocated in an attempt to consolidate into one section all of the requirements for emergency preparedness.

2015 CODE:

SECTION 403 ~~408~~
~~USE AND OCCUPANCY-RELATED~~ EMERGENCY PREPAREDNESS REQUIREMENTS

403.1 ~~408.1~~ General.   ~~In addition to the other requirements of this chapter, the provisions of this section are applicable to specific occupancies listed herein.~~ In addition to the requirements of Section 401, occupancies, uses and outdoor locations shall comply with the emergency preparedness requirements set forth in Sections 403.2 through 403.12.3.3. Where a fire safety and evacuation plan is required by Sections 403.2 through 403.11.4, evacuation drills shall be in accordance with Section 405 and employee training shall be in accordance with Section 406.

403.2 ~~408.2~~ Group A Occupancies.   An approved fire safety and evacuation plan in accordance with Section 404 shall be prepared and maintained for Group A occupancies, other than those occupancies used exclusively for purposes of religious worship with an occupant load less than 2,000, and for buildings containing both a Group A occupancy and an atrium. Group A occupancies shall comply with ~~the requirements of~~ Sections 403.2.1 through 403.2.4 ~~408.2.1 and 408.2.2 and Sections 401 through 406~~.

403.2.1 ~~408.2.1~~ Seating Plan.   In addition to the requirements of Section 404.2, t~~T~~he fire safety and evacuation plans for assembly occupancies shall include ~~the information required by Section 404.3 and~~ a detailed seating plan, occupant load and occupant load limit. Deviations from the approved plans shall be allowed provided the occupant load limit for the occupancy is not exceeded and the aisles and exit access-ways remain unobstructed.

403.2.2 ~~408.2.2~~ Announcements.   In theaters, motion picture theaters, auditoriums and similar assembly occupancies in Group A used for noncontinuous programs, an audible announcement shall be made not more than 10 minutes prior to the start of each program to notify the occupants of the location of the exits to be used in the event of a fire or other emergency.

> **Exception:**   In motion picture theaters, the announcement is allowed to be projected upon the screen in a manner approved by the fire code official.

403.2.3 Fire Watch Personnel.   Fire watch personnel shall be provided where required by Section 403.12.1.

FIRE SAFETY & EVACUATION PLANNING FOR BUSINESSES AND COVERED MALL BUILDINGS
April 2012

COLORADO SPRINGS FIRE DEPARTMENT
Division of the Fire Marshal

If you have any questions or comments regarding the information contained within, or if you need assistance interpreting these requirements, please contact:

Colorado Springs Fire Department
Division of the Fire Marshal
375 Printers Parkway
Colorado Springs, Colorado 80910
Tel: (719) 385-5978
Fax: (719) 385-7334

International Code Council®

Emergency preparedness, fire safety and evacuation plan

403.2.4 Crowd Managers.   Crowd managers shall be provided where required by Section 403.12.3.

403.3 Ambulatory Care Facilities.   Ambulatory care facilities shall comply with the requirements of Sections 403.3.1 through 403.3.3 as well as 401 and 404 through 406.

403.3.1 Fire Evacuation Plan.   The fire safety and evacuation plan required by Section 404 shall include a description of special staff actions. This shall include procedures for stabilizing patients in a defend-in-place response, staged evacuation, or full evacuation in conjunction with the entire building if part of a multitenant facility.

403 continues

403 continued

403.3.2 Fire Safety Plan. A copy of the plan shall be maintained at the facility at all times. Plan shall include all of the following in addition to the requirements of Section 404:

1. Locations of patients who are rendered incapable of self-preservation.
2. Maximum number of patients rendered incapable of self-preservation.
3. Area and extent of each ambulatory care facility.
4. Location of adjacent smoke compartments or refuge areas, where required.
5. Path of travel to adjacent smoke compartments.
6. Location of any special locking, delayed egress or access control arrangements.

403.3.3 Staff Training. Employees shall be periodically instructed and kept informed of their duties and responsibilities under the plan. Records of instruction shall be maintained. Such instruction shall be reviewed by the staff not less than every two months. A copy of the plan shall be readily available at all times within the facility.

403.3.4 Emergency Evacuation Drills. Emergency evacuation drills shall comply with Section 405. Emergency evacuation drills shall be conducted not less than four times per year.

Exceptions: The movement of patients to safe areas or to the exterior of the building is not required.

403.4 Group B Occupancies. An approved fire safety and evacuation plan in accordance with Section 404 shall be prepared and maintained for buildings containing a Group B occupancy where the Group B occupancy has an occupant load of 500 or more persons or more than 100 persons above or below the lowest level of exit discharge and for buildings having an ambulatory care facility.

403.5 Group E Occupancies. An approved fire safety and evacuation plan in accordance with Section 404 shall be prepared and maintained for Group E occupancies and for buildings containing both a Group E occupancy and an atrium. Group E occupancies shall also comply with Sections 403.5.1 through 403.5.3.

403.5.1 ~~408.3.1~~ First Emergency Evacuation Drill. The first emergency evacuation drill of each school year shall be conducted within 10 days of the beginning of classes.

~~**408.3.2 Emergency Evacuation Drill Deferral.** In severe climates, the *fire code official* shall have the authority to modify the emergency evacuation drill frequency specified in Section 405.2.~~

403.5.2 ~~408.3.3~~ Time of Day. Emergency evacuation drills shall be conducted at different hours of the day or evening, during the changing of classes, when the school is at assembly, during the recess or gymnastic

periods, or during other times to avoid distinction between drills and ac-
tual fires. ~~In Group R-2 college and university buildings, one required drill shall be held during hours after sunset or before sunrise.~~

403.5.3 ~~408.3.4~~ **Assembly Points.** Outdoor assembly areas shall be designated and shall be located a safe distance from the building being evacuated so as to avoid interference with fire department operations. The assembly areas shall be arranged to keep each class separate to provide accountability of all individuals.

403.6 Group F Occupancies. An approved fire safety and evacuation plan in accordance with Section 404 shall be prepared and maintained for buildings containing a Group F occupancy where the Group F occupancy has an occupant load of 500 or more persons or more than 100 persons above or below the lowest level of exit discharge.

403.7 Group H Occupancies. An approved fire safety and evacuation plan in accordance with Section 404 shall be prepared and maintained for Group H occupancies.

403.8 Group I Occupancies. An approved fire safety and evacuation plan in accordance with Section 404 shall be prepared and maintained for Group I occupancies. Group I occupancies shall also comply with Sections 403.8.1 through 403.8.3.

403.9 Group M Occupancies. An approved fire safety and evacuation plan in accordance with Section 404 shall be prepared and maintained for buildings containing a Group M occupancy, where the Group M occupancy has an occupant load of 500 or more persons or more than 100 persons above or below the lowest level of exit discharge, and for buildings containing both a Group M occupancy and an atrium.

403.10 Group R Occupancies. Group R occupancies shall comply with Sections 403.10.1 through 403.10.3.6.

Section 403 has been entirely rewritten. For brevity and clarity, only portions of the code text are shown. See the 2015 IFC for the complete code text.

Section 201.3 Terms Defined in Other Codes. The following is from the *International Building Code:*

Defend-in-Place. A method of emergency response that engages building components and trained staff to provide occupant safety during an emergency. Emergency response involves remaining in place, relocating within the building, or both, without evacuating the building.

CHANGE SIGNIFICANCE: Today's headlines and the tragic events that continue to unfold around the United States make this section more valuable today than ever before. Schools and some hotel chains are very diligent about emergency planning and preparedness. However, it is of increasing importance for the fire code official to enforce the provisions of this section.

403 continues

403 continued

The 2012 code splits such requirements between Section 404.2 and Section 408, making the code difficult to follow and apply. This change restructures Chapter 4 to place all of the core requirements in the front of the chapter in Section 403, which have been merged by occupancy classification or as otherwise appropriate.

This change is an editorial revision. Provisions have been relocated and text has been edited in an effort to improve readability and to clarify what is believed to be the current intent without technical change. One section dealing with maintenance of unoccupied tenant spaces in malls has been determined to be improperly located in Chapter 4 and has been moved to Chapter 3 with other vacant-use regulations.

Language regarding the defend-in-place or protect-in-place methods is now incorporated into Section 403. This concept has long been employed as the preferred method of fire response in hospitals due to the fragile nature of the occupants. Occupants in this setting are often dependent on the building infrastructure, and immediate evacuation would place their lives at risk. This infrastructure typically includes life-support systems such as medical gases, emergency power, and environmental controls that rely on continued building operation. This change identifies Group I-2 as a location where this type of emergency response is permitted.

PART 3

Building and Equipment Design Features

Chapters 5 through 19

International Fire Code (IFC) Part III contains requirements that provide fire fighters with a means of accessing a building and establishing a fire protection water supply for that building. Chapter 5 contains requirements for address numbers on buildings and provisions for fire department access roadways. Chapter 6 contains requirements for building systems, such as elevators, standby and emergency power systems (EPS), stationary battery systems and refrigeration systems. When specified by the *International Building Code* (IBC), buildings constructed using fire-resistive materials must be properly maintained to ensure the specified fire-resistance ratings are maintained. Chapter 7 specifies the requirements for maintenance of fire-resistance-rated construction.

Interior finish and decorative materials or furnishings offer fuel contribution and surfaces through which a fire can spread and transport heat and smoke to other parts of a room or to other rooms. Chapter 8 contains the most current requirements for regulating wall and ceiling finishes, decorative materials and furnishings.

Fire protection systems are required in accordance with Chapter 9. Chapter 9 specifies the requirements for automatic sprinkler systems, alternative fire-extinguishing systems, fire alarm and detection systems, standpipes, portable fire extinguishers, emergency alarm systems, smoke and heat vents, and smoke control systems. For materials that can have a detonation or deflagration hazard, Chapter 9 specifies the requirements for explosion control systems.

Merriam-Webster's *New Collegiate Dictionary* defines egress as "a place or means of going out." In the event of a fire or an emergency that requires the occupants to safely exit a building, Chapter 10 establishes the minimum requirements for means of egress from buildings.

Chapter 11 sets forth retroactive construction requirements for existing buildings. These provisions establish minimum fire-resistance-rating requirements for shafts as well as minimum means of egress requirements in existing buildings. It also establishes retroactive requirements for the installation of automatic sprinkler systems and fire alarm and detection systems in existing buildings or occupancies. ■

604.1

Emergency and Standby Power Systems

CHANGE TYPE: Addition

CHANGE SUMMARY: This change brings additional requirements related to emergency and standby power systems from the IBC into the IFC to provide for consistency and uniform enforcement. Load-transfer timing and duration are both quantified to assist the fire code official. Criteria have been added for Group I-2 occupancies that are located in flood hazard areas.

2015 CODE: 604.1 ~~Installation~~ General.   Emergency power systems and standby power systems required by this code or the *International Building Code* ~~shall be installed in accordance with this code, NFPA 110 and~~

Power switching (load transfer) system in hospital

~~NFPA 111. Existing installations shall be maintained in accordance with the original approval.~~ shall comply with Sections 604.1.1 through 604.1.8.

604.1.1 Stationary Generators. Stationary emergency and standby power generators required by this code shall be *listed* in accordance with UL 2200.

604.1.2 Installation. Emergency power systems and standby power systems shall be installed in accordance with *International Building Code*, NFPA 70, NFPA 110 and NFPA 111.

604.1.3 Load Transfer. Emergency power systems shall automatically provide secondary power within 10 seconds after primary power is lost, unless specified otherwise in this code. Standby power systems shall automatically provide secondary power within 60 seconds after primary power is lost unless specified otherwise in this code.

604.1.4 Load Duration. Emergency power systems and standby power systems shall be designed to provide the required power for a minimum duration of 2 hours without being refueled or recharged, unless specified otherwise in this code.

604.1.5 Uninterruptable Power Source. An uninterrupted source of power shall be provided for equipment when required by the manufacturer's instructions, the listing, this code, or applicable referenced standards.

604.1.6 Interchangeability. Emergency power systems shall be an acceptable alternative for installations that require standby power systems.

604.1.7 Group I-2 Occupancies. In Group I-2 occupancies, where an essential electrical system is located in flood hazard areas established in Section 1612.3 of the *International Building Code* and where new or replacement essential electrical system generators are installed, the system shall be located and installed in accordance with ASCE 24.

604.1.8 Maintenance. Existing installations shall be maintained in accordance with the original approval and Section 604.4.

SECTION 202
GENERAL DEFINITIONS

Emergency Power System. A source of automatic electric power of a required capacity and duration to operate required life safety, fire alarm, and detection and ventilation systems in the event of a failure of the primary power. Emergency power systems are required for electrical loads where interruption of the primary power could result in loss of human life or serious injuries.

Standby Power System. A source of automatic electric power of a required capacity and duration to operate required building, hazardous materials or ventilation systems in the event of a failure of the primary

604.1 continues

604.1 continued

<u>power. Standby power systems are required for electrical loads where interruption of the primary power could create hazards or hamper rescue or fire-fighting operations.</u>

CHANGE SIGNIFICANCE: The expansion of the existing Section 604.1 is intended to assist the fire code official by providing details such as installation requirements, load-transfer timing, load duration of the fuel supply, criteria for new or replacement installations for Group I-2 occupancies and maintenance requirements in one code section. Previously, this information was contained only in Section 2701 of the IBC. Definitions for emergency power systems (EPS) and standby power systems (SPS) have been added.

EPS and SPS are required based on a building's occupancy classification, height or depth of the building, building area, occupant load or specific hazardous materials or hazards to emergency responders. These systems are designed to provide electrical power to specified loads such as means-of-egress illumination, smoke control systems, emergency voice/alarm communication systems, fire service access elevators or occupant evacuation elevators.

This section now provides requirements for maximum load-transfer times. EPS must automatically transfer loads within 10 seconds after primary power is lost, and SPS must automatically transfer loads within 60 seconds after primary power is lost. Additionally, the amount of fuel needed for EPS and SPS is 2 hours minimum unless an alternative duration is required.

In recognition of the problems created when generators are rendered unusable due to flooding, this section now requires that new and replacement essential electrical systems in Group I-2 occupancies shall be installed in accordance with ASCE 24, Flood Resistant Design and Construction (American Society of Civil Engineers, Standard 24). ASCE 24 is a referenced standard in the *International Building Code*. Any building or structure that falls within the scope of the IBC that is proposed in a flood hazard area is to be designed in accordance with ASCE 24.

ASCE 24 tells the designer the minimum requirements and expected performance for the design and construction of buildings and structures in flood hazard areas. It is not a restatement of all of the National Flood Insurance Program regulations, but offers additional specificity, some additional requirements and some limitations. Buildings designed according to ASCE 24 are better able to resist flood loads and flood damage.

CHANGE TYPE: Addition

CHANGE SUMMARY: Essential electrical systems must comply with IBC Chapter 27 and NFPA 99. This change provides a clear path for the designer to know which standards apply when he or she is designing an essential electrical system for a Group I-2 occupancy.

2015 CODE: **604.2.6 Group I-2 Occupancies.**   Essential electrical systems for Group I-2 occupancies shall be in accordance with Section 407.10 of the *International Building Code.*

604.2.6, IBC 407.10 continues

604.2.6, IBC 407.10

Emergency and Standby Power Systems—Group I-2 Occupancies

Emergency generator

Hospital

604.2.6, IBC 407.10 continued

IBC Section 407.10 Electrical Systems. In Group I-2 occupancies, the essential electrical system for electrical components, equipment and systems shall be designed and constructed in accordance with the provisions of Chapter 27 and NFPA 99.

CHANGE SIGNIFICANCE: Currently, emergency power systems (EPS) are required to comply with NFPA 99 by the Center for Medicare/Medicaid Services (CMS) in order for a facility to receive federal reimbursement funds. By requiring compliance with NFPA 99 *Health Care Facilities Code*, the code official will ensure that the required power system is provided in Group I-2 occupancies. While there is a reference to NFPA 99 in NFPA 70 *National Electrical Code*®, there is no direct reference. This change closes a gap in the requirements. A reference to Chapter 27 comprehensively addresses electrical systems including references to NFPA 70; NFPA 110, Standard for Emergency and Standby Power Systems; and NFPA 111, Standard on Stored Electrical Energy Emergency and Standby Power Systems.

CHANGE TYPE: Clarification

CHANGE SUMMARY: The requirements for solar PV systems have been clarified and coordinated with the IBC and NFPA 70.

2015 CODE: 605.11 Solar Photovoltaic Power Systems. Solar photovoltaic (PV) power systems shall be installed in accordance with Sections 605.11.1 through ~~605.11.4~~ 605.11.2, the *International Building Code or International Residential Code,* and NFPA 70.

> **Exception:** ~~Detached, nonhabitable Group U structures including, but not limited to, parking shade structures, carports, solar trellises and similar structures shall not be subject to the requirements of this section.~~

605.11.1 Marking. ~~Marking is required on interior and exterior direct-current (DC) conduit, enclosures, raceways, cable assemblies, junction boxes, combiner boxes and disconnects.~~

605.11.1.1 Materials. ~~The materials used for marking shall be reflective, weather resistant and suitable for the environment. Marking as required in Sections 605.11.1.2 through 605.11.1.4 shall have all letters capitalized with a minimum height of 3/8 inch (9.5 mm) white on red background.~~

605.11 continues

605.11
Solar Photovoltaic Power Systems

Solar panel array on commercial building with access around perimeter and access to smoke and heat vents

605.11 continued

605.11.1.2 Marking Content. ~~The marking shall contain the words "WARNING: PHOTOVOLTAIC POWER SOURCE."~~

605.11.1.3 Main Service Disconnect. ~~The marking shall be placed adjacent to the main service disconnect in a location clearly visible from the location where the disconnect is operated.~~

605.11.1.4 Location of Marking. ~~Marking shall be placed on interior and exterior DC conduit, raceways, enclosures and cable assemblies every 10 feet (3048 mm), within 1 foot (305 mm) of turns or bends and within 1 foot (305 mm) above and below penetrations of roof/ceiling assemblies, walls or barriers.~~

605.11.2 Locations of DC Conductors. ~~Conduit, wiring systems, and raceways for photovoltaic circuits shall be located as close as possible to the ridge or hip or valley and from the hip or valley as directly as possible to an outside wall to reduce trip hazards and maximize ventilation opportunities. Conduit runs between sub arrays and to DC combiner boxes shall be installed in a manner that minimizes the total amount of conduit on the roof by taking the shortest path from the array to the DC combiner box. The DC combiner boxes shall be located such that conduit runs are minimized in the pathways between arrays. DC wiring shall be installed in metallic conduit or raceways when located within enclosed spaces in a building. Conduit shall run along the bottom of load bearing members.~~

~~605.11.3~~ 605.11.1 Access and Pathways. Roof access, pathways, and spacing requirements shall be provided in accordance with Sections ~~605.11.3.1 through 605.11.3.3.3~~ 605.11.1.1 through 605.11.1.3.3.

Exceptions:

1. ~~Residential structures shall be designed so that each photovoltaic array is no greater than 150 feet (45 720 mm) by 150 feet (45 720 mm) in either axis.~~ Detached, nonhabitable Group U structures including, but not limited to, parking shade structures, carports, solar trellises and similar structures.

2. ~~Panels/modules shall be permitted to be located up to the roof ridge~~ Roof access, pathways and spacing requirements need not be provided ~~where an alternative ventilation method approved by the fire chief has been provided or where~~ the fire chief has determined ~~vertical ventilation techniques~~ that rooftop operations will not be employed.

~~605.11.3.1~~ 605.11.1.1 Roof Access Points. Roof access points shall be located in areas that do not require the placement of ground ladders over openings such as windows or doors, and located at strong points of building construction in locations where the access point does not conflict with overhead obstructions such as tree limbs, wires, or signs.

~~605.11.3.2~~ 605.11.1.2 Residential Solar Photovoltaic Systems for ~~One- and Two-Family Dwellings~~ Group R-3 Buildings. ~~Access to residential systems for one- and two-family dwellings shall be provided in accordance with Sections 605.11.3.2.1 through 605.11.3.2.4.~~

Solar photovoltaic systems for Group R-3 buildings shall comply with Sections 605.11.1.2.1 through 605.11.1.2.5.

> **Exception:** These requirements shall not apply to structures designed and constructed in accordance with the *International Residential Code.*

605.11.1.2.1 Size of Solar Photovoltaic Array. Each photovoltaic array shall be limited to 150 feet (45,720 mm) by 150 feet (45,720 mm). Multiple arrays shall be separated by a 3-foot-wide (914 mm) clear access pathway.

605.11.3.2.1 605.11.1.2.2 Residential Buildings with Hip Hip Roof Layouts. Panels/ and modules installed on ~~residential buildings~~ Group R-3 buildings with hip roof layouts shall be located in a manner that provides a 3-foot-wide (914 mm) clear access pathway from the eave to the ridge on each roof slope where panels/ and modules are located. The access pathway shall be located at a ~~structurally strong~~ location on the building capable of supporting the ~~live load of~~ firefighters accessing the roof.

> **Exception:** These requirements shall not apply to roofs with slopes of two units vertical in 12 units horizontal (2:12) or less.

605.11.3.2.2 605.11.1.2.3 Residential Buildings with a sSingle Ridge Roofs. Panels/ and modules installed on ~~residential buildings~~ Group R-3 buildings with a single ridge shall be located in a manner that provides two, 3-foot-wide (914 mm) access pathways from the eave to the ridge on each roof slope where panels/ and modules are located.

> **Exception:** This requirement shall not apply to roofs with slopes of two units vertical in 12 units horizontal (2:12) or less.

605.11.3.2.3 605.11.1.2.4 Residential Buildings with Roofs with Hips and Valleys. Panels/ and modules installed on ~~residential buildings~~ Group R-3 buildings with roof hips and valleys shall be located no closer than 18 inches (457 mm) to a hip or a valley where panels/modules are to be placed on both sides of a hip or valley. Where panels are to be located on only one side of a hip or valley that is of equal length, the panels shall be permitted to be placed directly adjacent to the hip or valley.

> **Exception:** These requirements shall not apply to roofs with slopes of two units vertical in 12 units horizontal (2:12) or less.

605.11.3.2.4 605.11.1.2.5 Residential Building Allowance for Smoke Ventilation Operations. Panels/ and modules installed on ~~residential buildings~~ Group R-3 buildings shall be located no ~~higher~~ less than 3 feet (914 mm) ~~below~~ from the ridge in order to allow for fire department smoke ventilation operations.

> **Exception:** Panels and modules shall be permitted to be located up to the roof ridge where an alternative ventilation method approved by the fire chief has been provided or where the fire chief has determined vertical ventilation techniques will not be employed.

605.11 continues

605.11 continued

605.11.3.3 605.11.1.3 Other than ~~Residential Buildings~~ Group R-3 Buildings. Access to systems for ~~occupancies~~ buildings other than ~~one- and two-family dwellings~~ those containing Group R-3 occupancies shall be provided in accordance with Sections ~~605.11.3.3.1 through 605.11.3.3.3~~ 605.11.1.3.1 through 605.11.1.3.3.

> **Exception:** Where it is determined by the fire code official that the roof configuration is similar to that of a ~~one- or two-family dwelling~~ Group R-3 building, the residential access and ventilation requirements in Sections 605.11.3.2.1 through 605.11.3.2.5 shall be permitted to be used.

605.11.3.3.1 605.11.1.3.1 Access. There shall be a minimum 6-foot-wide (1829 mm) clear perimeter around the edges of the roof.

> **Exception:** Where either axis of the building is 250 feet (76 200 mm) or less, ~~there~~ the clear perimeter around the edges of the roof shall be permitted to be reduced to a minimum 4-foot-wide (1290 mm) ~~clear perimeter around the edges of the roof~~.

605.11.3.3.2 605.11.1.3.2 Pathways. The solar installation shall be designed to provide designated pathways. The pathways shall meet the following requirements:

1. The pathway shall be over areas capable of supporting ~~the live load of~~ fire fighters accessing the roof.
2. The centerline axis pathways shall be provided in both axes of the roof. Centerline axis pathways shall run where the roof structure is capable of supporting ~~the live load of~~ fire fighters accessing the roof.
3. Pathways shall be a straight line not less than 4 feet (1290 mm) clear to ~~skylights~~ roof standpipes or ventilation hatches.
4. ~~Shall be a straight line not less than 4 feet (1290 mm) clear to roof standpipes.~~
5. 4. Pathways shall provide not less than 4 feet (1290 mm) clear around roof access hatch with at least one not less than 4 feet (1290 mm) clear pathway to parapet or roof edge.

605.11.3.3.3 605.11.1.3.3 Smoke Ventilation. The solar installation shall be designed to meet the following requirements:

1. Arrays shall not be greater than 150 feet (45 720 mm) by 150 feet (45 720 mm) in distance in either axis in order to create opportunities for fire department smoke ventilation operations.
2. Smoke ventilation options between array sections shall be one of the following:
   2.1. A pathway 8 feet (2438 mm) or greater in width.
   2.2. A 4-foot (1290 mm) or greater in width pathway and bordering roof skylights or gravity-operated drop-out smoke and heat vents on at least one side.
   2.3. A 4-foot (1290 mm) or greater in width pathway and bordering all sides of nongravity-operated drop-out smoke and heat vents.

~~2.3.~~ 2.4. A 4-foot (1290 mm) or greater in width pathway and bordering 4-foot by 8-foot (1290 mm by 2438 mm) "venting cutouts" every 20 feet (6096 mm) on alternating sides of the pathway.

~~605.11.4~~ 605.11.2 Ground-Mounted Photovoltaic Arrays. Ground-mounted photovoltaic arrays shall comply with Sections 605.11 ~~through 605.11.2~~ and this section. Setback requirements shall not apply to ground-mounted, free-standing photovoltaic arrays. A clear, brush-free area of 10 feet (3048 mm) shall be required for ground-mounted photovoltaic arrays.

CHANGE SIGNIFICANCE: The revisions to Section 605.11 have reformatted and clarified the requirements for solar PV power systems. Table 605-1 is a cross-reference of the requirements in the 2012 IFC and the 2015 IFC.

One of the basic changes deals with the scope of the section. The previous exception for Group U occupancies has been relocated to Section 605.11.1. This relocation now only exempts Group U occupancies from accessways and pathways on the roof. Solar PV arrays on Group U occupancies must comply with the remainder of Section 605.11.

The requirements for marking and labeling have been removed from the IFC and are now found in Article 690.31 of the NFPA 70 *National Electrical Code.* This should not be a problem since a construction permit is required from the fire code official and the building code official with regard to electrical and structural requirements. The fire code official could refer to NFPA 70 since it is referenced in the IFC, or he or she could coordinate with the building official to ensure that the needed safeguards and signs are installed.

Many of the revisions to the access and pathway requirements are the result of a consensus process established by the Solar Energy Industries Association's (SEIA) Codes and Standards Working Group. Established in 1974, SEIA is the national trade association of the U.S. solar energy industry.

A new exception to Section 605.11.1.2 clarifies that the solar PV requirements in the IFC do not apply to one- and two-family dwellings constructed under the IRC. The IRC provides all construction requirements for the dwelling; the IFC regulates outside the building. The applicable IFC requirements for a building built under the IRC are fire fighter access, water supply, operations on the property, storage on the property, etc. This concept is repeated in Sections 605.11.1.2.2 through 605.11.1.3 where the terminology has changed from "residential building" to "Group R-3 building."

Section 605.11.1.2.1 applies to Group R-3 occupancies only and limits the size of the solar PV panel array to 150 feet in either axis, with separation of 3 feet between multiple arrays.

Exception 2 to Section 605.11.1 has been revised to simplify its application to a decision by the fire chief as to whether the fire department will or will not conduct rooftop operations. If the decision is that rooftop operations are not anticipated, requirements for spacing from hips, valleys, eaves, and ridges do not apply.

605.11 continues

605.11 continued

TABLE 605-1   Cross-reference of Section 605.11 in the 2012 IFC and 2015 IFC

2012 International Fire Code	2015 International Fire Code
605.11 Solar photovoltaic power systems	605.11 Solar photovoltaic power systems
Exception	*Moved to Section 605.11.3 Exception*
605.11.1 Marking	*Deleted – addressed in NFPA 70*
605.11.1.1 Materials	*Deleted – addressed in NFPA 70*
605.11.1.2 Marking content	*Deleted – addressed in NFPA 70*
605.11.1.3 Main service disconnect	*Deleted – addressed in NFPA 70*
605.11.1.4 Location of marking	*Deleted – addressed in NFPA 70*
605.11.2 Locations of DC conductors	*Deleted – addressed in NFPA 70*
605.11.3 Access and pathways	605.11.1 Access and pathways
Exception 1	*Moved to Section 605.11.1.2.1*
Exception 2	*Moved to Section 605.11.1.2.5 Exception*
605.11.3.1 Roof access points	605.11.1.1 Roof access points
605.11.3.2 Residential systems for one- and two-family dwellings	605.11.1.2 Solar photovoltaic systems for Group R-3 Buildings
-	Exception *(new exception)*
605.11.3.2.1 Residential buildings with hip roof layouts	605.11.1.2.2 Hip roof layouts
Exception	Exception
605.11.3.2.2 Residential buildings with a single ridge	605.11.1.2.3 Single-ridge roofs
Exception	Exception
605.11.3.2.3 Residential buildings with roof hips and valleys	605.11.1.2.4 Roofs with hips and valleys
Exception	Exception
605.11.3.2.4 Residential building smoke ventilation	605.11.1.2.5 Allowance for smoke ventilation operations
605.11.3.3 Other than residential buildings	605.11.1.3 Other than Group R-3 Buildings
Exception	Exception
605.11.3.3.1 Access	605.11.1.3.1 Access
Exception	Exception
605.11.3.3.2 Pathways	605.11.1.3.2 Pathways
Item 1	Item 1
Item 2	Item 2
Item 3	Item 3
Item 4	*Combined with Section 605.11.1.3.2, Item 3*
Item 5	Item 4
605.11.3.3.3 Smoke ventilation	605.11.1.3.3 Smoke ventilation
Item 1	Item 1
Item 2	Item 2
Item 2.1	Item 2.1
Item 2.2	Item 2.2
-	Item 2.3 *(new requirement)*
Item 2.3	Item 2.4
605.11.4 Ground-mounted photovoltaic arrays	605.11.2 Ground-mounted photovoltaic arrays

Section 605.11.1.3.3 has been revised to clarify access to and around smoke and heat vents and provide roof areas for smoke ventilation. Under Item 2, the designer has a choice of four methods to provide adequate access.

1. Item 2.1 requires pathways at least 8 feet in width.

2. Item 2.2 requires an accessway at least 4 feet in width to be provided to all skylights and gravity-operated drop-out smoke and heat vents. There must be an accessway bordering at least one side of every skylight and gravity-operated drop-out smoke and heat vent.

3. New Item 2.3 specifies that when nongravity-operated smoke and heat vents are used, an accessway at least 4 feet in width must lead to each vent and encircle it.

Section 605.11.2 has been revised to eliminate the reference to all of the requirements in Section 605.11. Now ground-mounted solar PV arrays are addressed as follows:

- Section 605.11 requires compliance with the IBC and NFPA 70. The IRC would not apply because the IRC covers construction of the structure only, and the ground-mounted device is not considered part of the structure.

- Section 605.11.2 specifies that setback requirements do not apply to ground-mounted solar PV arrays. This would include separation to property line or buildings in the IBC.

- Section 605.11.2 requires a clear area of 10 feet around the perimeter of the ground-mounted solar PV array. This is intended to prevent any electrical short or other electrical problem from igniting a fire in the grass and brush.

606.12

Pressure Relief Devices for Mechanical Refrigeration

CHANGE TYPE: Modification

CHANGE SUMMARY: The revisions to Section 606.12 clarify the code requirements and add references to two International Institute of Ammonia Refrigeration (IIAR) standards and one American Society of Heating, Refrigerating and Air-Conditioning Engineers, Inc. (ASHRAE) standard for design and operation of ammonia refrigeration systems.

2015 CODE: 606.12 <u>Discharge and Termination of Pressure Relief Devices and Purge Systems</u>.   Pressure relief devices, fusible plugs and purge systems <u>discharging to the atmosphere from</u> ~~for~~ refrigeration systems containing ~~more than 6.6 pounds (3 kg) of~~ flammable, toxic or highly toxic refrigerants <u>or ammonia</u> shall ~~be provided with an~~ *approved* ~~discharge system as required by~~ comply with Sections ~~606.12.1, 606.12.2 and 606.12.3~~ <u>606.12.3 through 606.12.5</u>. ~~Discharge piping and devices connected to the discharge side of a fusible plug or rupture member shall have provisions to prevent plugging the pipe in the event of the fusible plug or rupture member functions.~~

606.12.1 Standards.   <u>Refrigeration systems and the buildings in which such systems are installed shall be in accordance with ASHRAE 15.</u>

606.12.1.1 Ammonia Refrigeration.   <u>Refrigeration systems using ammonia refrigerant and the buildings in which such systems are installed shall comply with IIAR-2 for system design and installation and IIAR-7 for operating procedures.</u>

Screw compressors in a typical ammonia refrigeration machinery room *(Photo courtesy of the International Institute of Ammonia Refrigeration)*

Pressure relief valves on an ammonia refrigeration system *(Photo courtesy of the International Institute of Ammonia Refrigeration)*

606.12.2 Fusible Plugs and Rupture Members. Discharge piping and devices connected to the discharge side of a fusible plug or rupture member shall have provisions to prevent plugging the pipe in the event the fusible plug or rupture member functions.

606.12.1 606.12.3 Flammable Refrigerants. Systems containing more than 6.6 pounds (3 kg) of flammable refrigerants having a density equal to or greater than the density of air shall discharge vapor to the atmosphere only through an approved treatment system in accordance with Section 606.12.4 606.12.6 or a flaring system in accordance with Section 606.12.5 606.12.7. Systems containing more than 6.6 pounds (3 kg) of flammable refrigerants having a density less than the density of air shall be permitted to discharge vapor to the atmosphere provided that the point of discharge is located outside of the structure at not less than 15 feet (4,572 mm) above the adjoining grade level and not less than 20 feet (6,096 mm) from any window, ventilation opening or *exit*.

606.12.2 606.12.4 Toxic and Highly Toxic Refrigerants. Systems containing more than 6.6 pounds (3 kg) of toxic or highly toxic refrigerants shall discharge vapor to the atmosphere only through an approved treatment system in accordance with Section 606.12.4 606.12.6 or a flaring system in accordance with Section 606.12.5 606.12.7.

606.12.3 606.12.5 Ammonia Refrigerant. Systems containing more than 6.6 pounds (3 kg) of ammonia refrigerant shall discharge vapor to the atmosphere in accordance with one of the following methods: through an approved treatment system in accordance with Section 606.12.4, a flaring system in accordance with Section 606.12.5, or through an approved ammonia diffusion system in accordance with Section 606.12.6, or by other approved means.

1. Directly to atmosphere when the fire code official determines, on review of an engineering analysis prepared in accordance with Section 104.7.2, that a fire, health or environmental hazard would not result from atmospheric discharge of ammonia.
2. Through an approved treatment system in accordance with Section 606.12.6.
3. Through a flaring system in accordance with Section 606.12.7.
4. Through an approved ammonia diffusion system in accordance with Section 606.12.8.
5. By other approved means.

Exceptions:

1. Ammonia/water absorption systems containing less than 22 pounds (10 kg) of ammonia and for which the ammonia circuit is located entirely outdoors.
2. When the fire code official determines, on review of an engineering analysis prepared in accordance with Section 104.7.2, that a fire, health or environmental hazard would not result from discharging ammonia directly to the atmosphere.

606.12 continues

606.12 continued

606.12.4 606.12.6 Treatment Systems. Treatment systems shall be designed to reduce the allowable discharge concentration of the refrigerant gas to not more than 50 percent of the IDLH at the point of exhaust. Treatment systems shall be in accordance with Chapter 60.

606.12.5 606.12.7 Flaring Systems. Flaring systems for incineration of flammable refrigerants shall be designed to incinerate the entire discharge. The products of refrigerant incineration shall not pose health or environmental hazards. Incineration shall be automatic upon initiation of discharge, shall be designed to prevent blowback and shall not expose structures or materials to threat of fire. Standby fuel, such as LP gas, and standby power shall have the capacity to operate for one and one-half the required time for complete incineration of refrigerant in the system. Standby electrical power, where required to complete the incineration process, shall be in accordance with Section 604.

606.12.6 606.12.7 Ammonia Diffusion Systems. Ammonia diffusion systems shall include a tank containing 1 gallon of water for each pound of ammonia (8.3 L of water for each 1 kg of ammonia) that will be released in 1 hour from the largest relief device connected to the discharge pipe. The water shall be prevented from freezing. The discharge pipe from the pressure relief device shall distribute ammonia in the bottom of the tank, but no lower than 33 feet (10,058 mm) below the maximum liquid level. The tank shall contain the volume of water and ammonia without overflowing.

CHAPTER 80
REFERENCED STANDARDS

ASHRAE

American Society of Heating, Refrigerating and Air-Conditioning Engineers, Inc.
    1791 Tullie Circle, NE
    Atlanta, GA 30329
    15-2013 Safety Standard for Refrigeration Systems 606.12.1

IIAR

International Institute of Ammonia Refrigeration
    1001 N. Fairfax Street, Suite 503
    Alexandria, VA 22314
    IIAR-2-2014 Equipment, Design, and Installation of Closed-Circuit Ammonia Mechanical Refrigerating Systems 606.12.1.1
    IIAR-7-2013 Developing Operating Procedures for Closed-Circuit Ammonia Mechanical Refrigerating 606.12.1.1

CHANGE SIGNIFICANCE: The revisions to Section 606.12 simplify the section through reorganization and by relocating some of the provisions to subsections.

The addition of "discharging to atmosphere" is consistent with the text in the sections governing flammable refrigerants (Section 606.12.3), toxic/highly toxic refrigerants (Section 606.12.4), and ammonia refrigerant (Section 606.12.5). This is intended to make it clear that restrictions on vent termination do not, and never did, apply to relief arrangements that are internal to a system (i.e., not routed to atmosphere).

References have been added to ASHRAE 15, Safety Standard for Refrigeration Systems, IIAR-2-2014, Equipment, Design, and Installation of Closed-Circuit Ammonia Mechanical Refrigerating Systems and IIAR-7-2013, Developing Operating Procedures for Closed-Circuit Ammonia Mechanical Refrigerating. IIAR-2 includes requirements for refrigerant-leak detection alarms. IIAR-7 establishes a minimum standard of mandatory operating procedures for ammonia refrigeration systems.

Other modifications in the subsections of Section 606.12 include requirements for discharge piping and devices that apply to fusible plugs and rupture members for *all* refrigeration systems.

The limitation to 6.6 pounds applied in the 2012 IFC for systems containing flammable refrigerants, but the threshold was in Section 606.12, which limited the requirements in the entire section to mechanical systems over 6.6 pounds. With the relocation of the size limitation, all of the sections apply to all systems, but the requirement for a treatment or flaring system apply only when a system containing flammable refrigerants exceeds 6.6 pounds.

Similarly, for refrigeration systems containing toxic or highly toxic refrigerants, a treatment system is required only when a system containing more than 6.6 pounds.

Requirements for refrigeration systems containing ammonia have been clarified. The capacity threshold of 6.6 pounds has also been added to this section providing consistency with the requirements for flammable, toxic or highly toxic refrigerants.

607.6

Protection of Fire Service Access Elevators and Occupant Evacuation Elevators

CHANGE TYPE: Addition

CHANGE SUMMARY: This is a new requirement to ensure that devices designed to prevent water from infiltrating into fire service access elevator hoistways and occupant evacuation elevator hoistways are properly maintained.

2015 CODE: <u>607.6 Water Protection of Hoistway Enclosures.</u> <u>Methods to prevent water from infiltrating into a hoistway enclosure required by Section 3007.4 and Section 3008.4 of the *International Building Code* shall be maintained.</u>

CHANGE SIGNIFICANCE: The *International Building Code* requires a method to prevent sprinkler water from penetrating a hoistway enclosure for both fire service access elevators (IBC Section 3007.4) and occupant evacuation elevators (IBC Section 3008.4). Table 607.6-1 compares the various features between occupant evacuation elevators and fire service access elevators.

The requirement for preventing water infiltration first appeared in the 2009 IBC for occupant evacuation elevators. The concern is to protect the reliability of the elevator operation and keep water out of the hoistway. The shunt-trip devices required by IBC Section 3005.5 are not installed in these particular elevators, so preventing water from affecting the braking system is critical. In the 2012 IBC, the same requirement was applied to fire service access elevators.

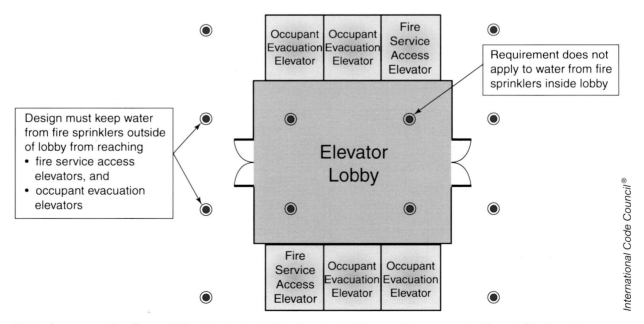

Water from operating fire sprinklers located outside elevator lobbies must be prevented from infiltrating into hoistways for fire service access elevators and occupant evacuation elevators.

International Code Council®

TABLE 607.6-1   Features of Occupant Evacuation Elevators and Fire Service Access Elevators

OCCUPANT EVACUATION ELEVATORS	FIRE SERVICE ACCESS ELEVATORS
Design option for buildings taller than 420 feet in lieu of additional stairway	Minimum of two elevator cars required when an occupied floor in the building exceeds 120 feet above lowest level of fire department vehicle access
Designed for self-evacuation and egress of occupants who are not capable of traveling the stairs in high-rise buildings	Designed for use during fire service operations
Phase I and II fire operations	Phase I and II fire operations
Fire sprinklers are prohibited in the elevator shaft and the elevator machine rooms	Fire sprinklers are prohibited in the elevator shaft and the elevator machine rooms
Shunt trip for elevator shutdown is prohibited	Shunt trip for elevator shutdown is prohibited
Elevator lobby with a minimum of 1-hour construction	Elevator lobby with a minimum of 1-hour construction
Lobby doors, other than the elevator car doors, must have a minimum of ¾-hour fire-protection rating with a vision panel	Lobby doors, other than the elevator car doors, must have a minimum of ¾-hour fire-protection rating
Lobby size must provide a minimum of 3 square feet/person based on 25% of the occupant load of the story, PLUS 1 wheelchair space (30″ × 48″) for each 50 persons based on the occupant load of the story	Lobby size must be a minimum of 150 square feet with a minimum dimension of 8 feet
Two-way communication from the elevator lobby to the fire command center	Not required
Elevator operation monitored at the fire command center	Elevator operation monitored at the fire command center
Interior exit stairway directly accessible from elevator lobby	Interior exit stairway directly accessible from elevator lobby
Building must be equipped with an emergency voice/alarm communication system	Not required
Standby power required for elevator equipment and HVAC in elevator machine room	Standby power required for elevator equipment, HVAC in elevator machine room and hoistway lighting
Method to prevent water discharged from fire sprinklers located outside the lobby from infiltrating the hoistway	Method to prevent water discharged from fire sprinklers located outside the lobby from infiltrating the hoistway

It is important to stress three items:

1. The requirement for preventing water from entering the hoistway does not apply to all elevators. It is only applicable to fire service access elevators and occupant evacuation elevators.

2. The source of water that must be addressed is from the fire sprinkler system and not from firefighter hoses.

3. The water of concern is limited to sprinkler activations outside the lobby.

The actual method of preventing water from entering the hoistway is not specified; it is a performance-based requirement. Prevention could be accomplished by trench drains in the floor, a slight slope in the floor as it approaches the hoistway, curbs, or gasketed openings.

Whichever method is selected, the dependability of the system to not allow water into the hoistway must be maintained. If a drain system is installed, drains can become clogged by dirt and debris; if gaskets are used, they can wear and deteriorate. It is important that the integrity of these systems be maintained so that the elevators remain safe during their use in a fire.

609.2

Commercial Cooking Appliances Producing a Low Volume of Grease-laden Vapors

CHANGE TYPE: Modification

CHANGE SUMMARY: Type I exhaust hoods are not required over electric cooking appliances when the appliances produce a minimal amount of grease-laden vapors.

2015 CODE: 609.2 Where Required. A Type I hood shall be installed at or above all commercial cooking appliances and domestic cooking appliances used for commercial purposes that produce grease vapors.

> **Exception:** A Type I hood shall not be required for an electric cooking appliance where an approved testing agency provides documentation that the appliance effluent contains 5 mg/m³ or less of grease when tested at an exhaust flow rate of 500 cfm (0.236 m³/s) in accordance with UL 710B.

CHANGE SIGNIFICANCE: This proposal brings consistency between the IFC and the IMC.

There are many situations where the amount of grease-laden vapor is very low to almost non-existent and a Type I hood is not needed. A test method is specified to determine when a Type I hood is not required.

The exception refers to a test in UL 710B, Standard for Recirculating Systems. The exception does not require an appliance listed to UL 710B; it requires only that the cooking appliance has been evaluated using the specific test method in UL 710B. The particular test is the "Emissions Test,"

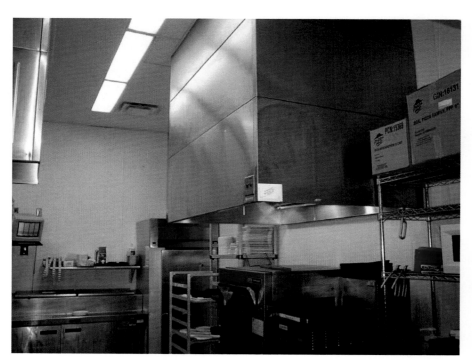

A Type I exhaust hood is not required for this conveyor pizza oven since it has been tested to verify that it produces below the threshold rate for grease-laden vapors. A Type II exhaust hood is still required. *(Photo courtesy of Target Corporation, Minneapolis, MN)*

which measures the amount of grease-laden effluent discharged from the appliance over a continuous 8-hour test cooking period. This test method is derived from the U.S. Environmental Protection Agency *Test Method 202, Determination of Condensable Particulate Emissions from Stationary Sources.* Essentially, samples are selected, prepared, and tested, and the effluent is captured. Then the effluent is analyzed to determine how much grease was contained in the effluent. If the effluent contained 5 mg/m^3 or less of grease, then the exception would apply.

There are cooking appliances that have been tested to the Emissions Test that do not need a Type I hood, because there is insufficient quantity of grease-laden vapors produced during the cooking operation. An example would be a steam table.

Section 904.2.2 states that a fire-extinguishing system must be provided over cooking appliances that are required to have a Type I hood. Therefore, if a Type I hood is not required because of the exception in Section 609.2, then a fire-extinguishing system is not required either. The IMC will still require a hood and exhaust system, but it can be a Type II hood.

609.3.3.2

Inspection and Cleaning of Commercial Kitchen Exhaust Hoods

Cleaning of commercial kitchen exhaust systems shall be in accordance with ANSI/IKECA C10

CHANGE TYPE: Modification

CHANGE SUMMARY: This section references a new standard that addresses the cleaning of commercial cooking exhaust hoods and ducts.

2015 CODE: 609.3.3.2 Grease Accumulation. If during the inspection it is found that hoods, grease-removal devices, fans, ducts, or other appurtenances have an accumulation of grease, such components shall be cleaned in accordance with ANSI/IKECA C10.

609.3.3.3 Records. Records for inspections shall state the individual and company performing the inspection, a description of the inspection and when the inspection took place. Records for cleanings shall state the individual and company performing the cleaning and when the cleaning took place. Such records shall be completed after each inspection or cleaning, and maintained on the premises for a minimum of three years and be copied to the *fire code official* upon request.

609.3.3.3.1 Tags. Where a commercial kitchen hood or duct system is inspected, a tag containing the service provider name, address, telephone number, and date of service shall be provided in a conspicuous location. Prior tags shall be covered or removed.

CHAPTER 80
REFERENCED STANDARDS

IKECA

International Kitchen Exhaust Cleaning Association
100 North 20th Street, Suite 400
Philadelphia, PA 19103
C10-2011 *Standard for Cleaning of Commercial Kitchen Exhaust Systems* 609.3.3.2

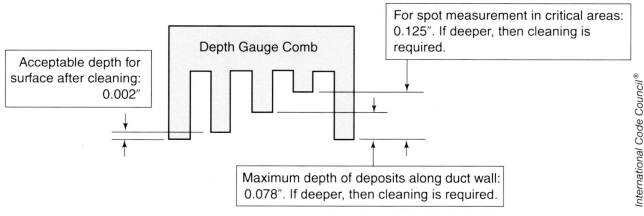

The depth of the grease accumulation is measured during the inspection to determine if cleaning is needed.

CHANGE SIGNIFICANCE: Commercial kitchen exhaust systems remove heat, smoke, fumes, steam and grease-laden vapors resulting from cooking operations. The interior walls of the exhaust systems become contaminated with grease and cooking by-products over time. Accumulations of these combustible contaminants create a fire hazard. Mitigation of this fire hazard requires periodic cleaning of commercial kitchen exhaust systems.

ANSI/IKECA C10 has been promulgated by the International Kitchen Exhaust Cleaning Association and provides guidance in determining the frequency and necessity for cleaning commercial kitchen exhaust systems. IFC Table 609.3.3.1 addresses the frequency of inspection. If that inspection identifies contamination, the commercial kitchen exhaust system must be cleaned. ANSI/IKECA C10 Section 9.1.3 refers to the use of a depth-gauge comb to measure the thickness of deposits on the duct walls. Duct cleaning is deemed necessary when the depth of the grease and deposits exceed 0.078 inches.

ANSI/IKECA C10 specifies acceptable methods for cleaning exhaust systems and components, and sets standards for acceptable cleanliness after the cleaning process is completed. The standard will provide consistency in the cleaning operation and thereby reduce the potential fire hazards associated with commercial kitchen exhaust systems.

Section 609.3.3.3 specifies that records must be maintained. The records include information on inspections of the exhaust system and information on cleaning operations of the exhaust system. These records must be maintained by the business.

Section 609.3.3.1 requires that when a hood or duct system is inspected, a tag shall be placed in a conspicuous location. This is similar to the inspection/service tag on portable fire extinguishers. The IFC requires this tag when the commercial kitchen exhaust system is inspected, and the tag shall be placed so that the most recent inspection is the only tag visible. This can be accomplished by either removing the previous tag or adhering the new tag to the equipment and placing it over the previous tag. If the inspection determines that cleaning is necessary, then the cleaning must be performed in accordance with ANSI/IKECA C10. ANSI/IKECA C10 Section 11.1 requires that a tag be affixed to the equipment after cleaning is completed, similar to the inspection tag.

609.4

Gas-fired Appliance Connections

CHANGE TYPE: Addition

CHANGE SUMMARY: Listed flexible connectors are required between the fixed fuel-gas piping and cooking appliances on casters or other appliances that are moved for cleaning.

2015 CODE: <u>**609.4 Appliance Connection to Building Piping.** Gas-fired commercial cooking appliances installed on casters and appliances that are moved for cleaning and sanitation purposes shall be connected to the piping system with an appliance connector listed as complying with ANSI Z21.69. The commercial cooking appliance connector installation shall be configured in accordance with the manufacturer's installation instructions. Movement of appliances with casters shall be limited by a restraining device installed in accordance with the connector and appliance manufacturer's instructions.</u>

CHAPTER 80
REFERENCED STANDARDS

ANSI

<u>Z21.69/CSA 616-09 *Connectors for Movable Gas Appliances*</u> 609.4

CHANGE SIGNIFICANCE: Typically, gas-fired appliances are connected to the building piping system with flexible connectors. The traditional flexible connector is designed to be manipulated and shaped to make the final connection. These traditional flexible connectors are not designed for repetitive movement. Repeated flexing causes weakening and tearing of the connector resulting in leaks and fires, since the ignition sources are so close.

The listed flexible connector (blue) allows for the repeated moving and repositioning of the cooking appliance resulting from routine cleaning. *(Photo courtesy of James Carver with El Segundo, CA, Fire Department)*

However, certain appliances are designed to be moved for routine maintenance and cleaning. This is typical of cooking appliances, where the appliance must be moved to clean the grease, cooking splatter, and debris that has accumulated near the appliance. This new section requires a specifically designed connector for appliances that are planned to be repeatedly moved. These connectors are tested to withstand this repeated movement without failure.

IFGC Section 411.1.1 also requires these listed connectors for cooking appliances that are on casters and appliances that are moved for cleaning. So at the time of initial construction, the IFGC requirements would be followed and inspected by the plumbing or mechanical inspector. However, typically when those appliances are replaced, permits are not obtained, and therefore, no specific inspections are made on the installations. Adding this requirement into the IFC provides the ability for the fire code official to inspect the installations and enforce these provisions.

The additional requirement for a restraining device is intended to not overextend the listed flexible connector and put a strain on the fittings.

611

Hyperbaric Facilities

CHANGE TYPE: Addition

CHANGE SUMMARY: A new Section 611 on hyperbaric facilities has been added to the IFC. According to the provisions of this section, these facilities shall be inspected, tested and maintained in accordance with NFPA 99. Records shall be kept and made available to the fire code official.

2015 CODE:

SECTION 611
HYPERBARIC FACILITIES

611.1 General. Hyperbaric facilities shall be inspected, tested and maintained, in accordance with NFPA 99.

611.2 Records. Records shall be maintained of all testing and repair conducted on the hyperbaric chamber and associated devices and equipment. Records shall be available to the fire code official.

CHANGE SIGNIFICANCE: The hyperbaric facilities referenced in the IFC use oxygen at pressures greater than normal atmosphere (sea level). The word "hyperbaric" is from the Greek root "hyper" meaning "over, above" and "baro" meaning "weight." Therefore, hyperbaric is "above the (normal) weight" of the atmosphere.

Hyperbaric medicine, also known as hyperbaric oxygen therapy (HBOT), is the medical use of oxygen at a level higher than atmospheric pressure. The equipment required consists of a pressure chamber, which may be of rigid or flexible construction, and a means of delivering 100 percent oxygen. The use of oxygen therapy and these chambers makes up one of the fastest-growing segments of the healthcare industry. The fire hazards of oxygen-enriched environments are well documented.

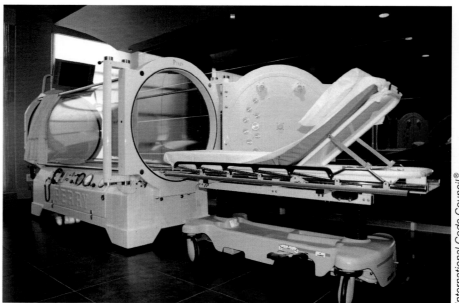

International Code Council®

Hyperbaric chamber

This change introduces hyperbaric facilities to the IFC. NFPA 99, *Health Care Facilities Code,* is the referenced document for inspection, testing and maintenance. Chapter 14 of NFPA 99 provides code requirements for these facilities. Fatalities have occurred in hyperbaric chambers as a result of fires within the chamber. Material that might not normally ignite in air at normal air pressure can ignite and burn quite vigorously when in a high oxygen concentration at elevated pressures. Materials that are not expected to ignite can become flammable when in a hyperbaric chamber. Materials such as certain flame-resistant fabrics, epoxy compounds and certain asbestos blankets have been known to burn vigorously in hyperbaric facilities.[1]

Another hazard associated with hyperbaric facilities is the fact that a hyperbaric chamber is a large pressure vessel. The sudden release of the pressure within the hyperbaric facility can result in damage to the structure and injury to adjacent personnel.

The code official's consultation with the safety director for the business operating the hyperbaric facility should occur during each inspection.

---

[1]NFPA 99 *Health Care Facilities Code*, NFPA, 2012 Edition, Section B.14.1.1.3.1, page 99–189.

807

Decorative Materials Other Than Decorative Vegetation in New and Existing Buildings

CHANGE TYPE: Modification

CHANGE SUMMARY: The requirements for decorative materials other than decorative vegetation have been reorganized and clarified.

2015 CODE:

SECTION 807
DECORATIVE MATERIALS OTHER THAN DECORATIVE VEGETATION IN NEW AND EXISTING BUILDINGS

Note: An outline of Section 807 is shown here. For the entire text please refer to the 2015 IFC.

807.1 General.
807.2 Limitations.
807.3 Combustible Decorative Materials.
807.4 Acceptance Criteria and Reports.
807.5 Occupancy-Based Requirements.
807.5.1 Group A.
807.5.1.1 Foam Plastics.
807.5.1.2 Motion Picture Screens.
807.5.1.3 Wood Use in Places of Religious Worship.
807.5.1.4 Pyroxylin Plastic.
807.5.2 Group E.
807.5.2.1 Storage in Corridors and Lobbies.
807.5.2.2 Artwork in Corridors.
807.5.2.3 Artwork in Classrooms.
807.5.3 Groups I-1 and I-2.
807.5.3.1 Group I-1 and Group I-2 Condition 1 within Units.
807.5.3.2 In Group I-1 and Group I-2 Condition 1 for Areas other than within Units.

Decorative combustible materials inside building

807.5.3.3 In Group I-2 Condition 2.
807.5.3.4 Other Areas in Groups I-1 and I-2.
807.5.4 Group I-3.
807.5.5 Group I-4.
807.5.5.1 Storage in Corridors and Lobbies.
807.5.5.2 Artwork in Corridors.
807.5.5.3 Artwork in Classrooms.
807.5.6 Dormitories in Group R-2.

CHANGE SIGNIFICANCE: The revision of Section 807 provides clarification to the regulation of combustible decorative materials allowed within a room or space. This section now contains provisions for "artwork" and "curtains, draperies, fabric hangings, and other similar combustible decorative materials."

Section 807.1 is now a general charging statement for Section 807. The provisions now clarify that the requirements in Section 807 do not apply to decorative vegetation.

Section 807.3 states that in all occupancies except Group I-3, curtains, draperies, fabric hangings, and similar combustible decorative materials are limited to 10 percent of the wall space or ceiling area to which the decorative materials are attached. There are three exceptions to this limitation of 10 percent.

1. Exception 1 increases the allowance to 75 percent in Group A auditoriums protected with a fire sprinkler system.

2. Exception 2 increases the allowance to 50 percent in Group R-2 dormitories protected with a fire sprinkler system. Under this reference to Section 903.3.1 for the fire sprinkler system design, the fire sprinkler system could comply with NFPA 13, 13R, or 13D.

3. Exception 3 is specific to fabric partitions suspended from the ceiling of a Group B or M occupancy. The exception states that if the partitions comply with Section 807.4, which means compliance with either NFPA 701 or NFPA 289, then there is no limit to their use.

Another key item in Section 807.3 is that fixed or movable walls and partitions, paneling, wall pads, and crash pads shall be regulated as interior finishes, not combustible decorative materials.

Section 807.4 provides for two test methods to be used when decorative materials are required to exhibit improved fire performance. In the 2012 IFC, some use groups required photographs and paintings to be tested and certified to NFPA 701, Standard Methods of Fire Tests for Flame-propagation of Textiles and Films, but the scope of NFPA 701 covers fabrics, curtains, draperies, and window treatments. NFPA 289, Standard Method of Fire Test for Individual Fuel Packages covers individual fuel packages and contains testing criteria for single decorative items, exhibit booths, and stage settings.

Section 807.5.2 is a new section that limits the quantity of artwork and teaching materials that can be displayed within classrooms. The limitation is to 50 percent of the wall space, based on the particular wall that the material is attached to.

807 continues

807 continued

Section 807.5.3.1 applies only to Group I-1 and Group I-2 Condition 1 and limits the quantity of combustible decorative material in sleeping units and dwelling units of sprinklered occupancies to 50 percent of the wall space, based on the particular wall that the material is attached to. This allowance is based on a fire sprinkler system design in accordance with NFPA 13.

Section 807.5.3.2 applies only to sprinklered Group I-1 and Group I-2 Condition 1 occupancies, but outside the sleeping units and dwelling units. The quantity of combustible decorative materials cannot exceed 30 percent of the wall space, based on the particular wall that the material is attached to.

Section 807.5.3.3 establishes a limit of 30 percent of the wall area in Group I-2 Condition 2 equipped with a fire sprinkler system.

Section 807.5.3.4 states that in Group I-2 Condition 2 occupancies that are not equipped with a fire sprinkler system, the allowable quantity of combustible decorative materials shall not create a hazard with regard to fire development or fire spread.

Section 807.5.5 applies to Group I-4 occupancies and limits artwork and teaching materials in a corridor to 20 percent of the wall space, and in a classroom to 50 percent of the wall space.

Section 807.5.6 applies to Group R-2 dormitories. Within sleeping units and dwelling units, the allowable quantity of combustible decorative materials shall not create a hazard with regard to fire development or fire spread.

901.4.1
Required Fire Protection Systems

CHANGE TYPE: Clarification

CHANGE SUMMARY: The code has been clarified concerning how an inspector can determine if a fire protection system is to be considered a "required" system or a "nonrequired" system.

2015 CODE: 901.4.1 Required Fire Protection Systems. Fire protection systems required by this code or the *International Building Code* shall be installed, repaired, operated, tested, and maintained in accordance with this code. <u>A fire protection system for which a design option, exception or reduction to the provisions of this code, or the *International Building Code* has been granted shall be considered to be a required system.</u>

CHANGE SIGNIFICANCE: The sentence added to Section 901.4.1 clarifies the code when an inspector is determining whether a fire protection system is a required system or a nonrequired system. This determination is important when the inspector is evaluating the application of Section 901.6.

IFC Section 901.6 contains requirements for required fire protection systems and nonrequired fire protection systems. Required fire protection systems must be maintained in an operative condition at all times. Nonrequired fire protection systems must be either

1. maintained in an operative condition at all times, or
2. removed.

The potential to remove a fire protection system has led to the need to clarify the intent of previous codes as to what is considered a required fire protection system. There are times when either the IBC or IFC do not require a fire protection system for a specific building, but either the owner or the designer has opted to install a fire protection system. A good example is the installation of an automatic sprinkler system that was not required by code but resulted in a reduction in property insurance premiums. Section 901.4.2 requires that such an automatic sprinkler system must comply with the code and the appropriate design standard, even though the system is not required by the code. Section 901.6 requires that the automatic sprinkler system must be maintained, tested, and inspected even though it was not a required system, or it can be removed.

If a building owner decides to remove the fire protection system rather than continue to maintain it, the code will allow this system to be removed provided that the system is nonrequired. This revision provides clarification to assist the code official in making the determination whether a fire protection system is a nonrequired or a required system.

Based on this section, some examples below demonstrate when a fire protection system will be considered a required system.

Example 1: A Group A-2 building is constructed, and the occupant load exceeds 100. IFC Section 903.2.1.2 requires an automatic sprinkler system.

Is that fire sprinkler part of a "required" fire protection system?

901.4.1 continues

901.4.1 continued

Example 2: A single-story Group B building is of Type VB construction and has a floor area of 25,000 square feet. IBC Table 503 allows only 9,000 square feet for this building, but IBC Section 506.3 allows an increase to 36,000 square feet when an automatic sprinkler system is installed. This system is a required system.

Example 3: A Group B building is constructed with an occupant load 40. The common path of egress travel is 90 feet. IBC/IFC Table 1006.2.1 limits the common path of travel to 100 feet when the building is protected with an automatic sprinkler system, but only 75 without the sprinkler system. The owner has a choice: install either a second exit or a sprinkler system. The owner chooses to install the automatic sprinkler system. This system is a required system.

　If the building design takes advantage of any of the modifications or reductions that are allowed by the code, then the fire protection system becomes a required system.

Example 4: The owner of the Group B building in Example 3 chose to install a second exit *and* install a sprinkler system, since the last building he had burned down. None of the modifications in either the IBC or the IFC were utilized in the design of the building. However, 8 years after construction, the owner decided to change the interior wall finishes in the exit access corridor and install a product with a flame spread rating of Class C. Table 803.3 allows Class C interior finish materials when protected with a fire sprinkler system. This sprinkler system is now considered a required system.

Example 5: A sprinklered Group F-1 building is proposed for construction. IBC/IFC Table 1017.2 limits the exit access travel distance to 250 feet. An alternate method is requested by the owner and approved by the code official to increase the travel distance to 275 feet based on the installation of an automatic fire alarm system. The fire alarm system is not required by code, but based on the alternate method it is now considered a required system.

Example 6: An existing restaurant classified as a Group A-2 occupancy is being remodeled to a retail stationery store and will receive a Certificate of Occupancy as a Group M occupancy. The previous restaurant was provided with a fire sprinkler system and a fire alarm system since the occupant load was 400 persons. The new Group M will have an occupant load of 60 persons. The sprinkler system and the fire alarm system are not required in the Group M occupancy. The owner can opt to either maintain the systems or remove them. If the owner removes the systems, all visible components must be removed. This would include the following items:

- Fire sprinklers
- Fire department connection
- Fire sprinkler riser
- Any exposed fire sprinkler system piping
- Fire alarm control panel
- Manual fire alarm boxes
- Audible and visual notification appliances

- Smoke detectors, including the device over the fire alarm control panel
- All exposed wiring and components visible from the occupied areas
- All wiring and components located above the ceiling, if the concealed space is used as a plenum. (See *Significant Changes to the International Fire Code, 2015 Edition*, Sections 315.6 and 605.12.)

Modifications, exceptions, reductions or other design options are found in the IBC and the IFC. If just a single building component is altered because of the fire protection system, then this system is considered a required one. As a required system, the option to remove the system no longer applies. This system must be maintained in an operable condition at all times.

901.8.2

Removal of Existing Occupant-Use Hose Lines

CHANGE TYPE: Addition

CHANGE SUMMARY: Existing 1½-inch hose lines can be removed under certain circumstances.

2015 CODE: **901.8.2 Removal of Existing Occupant-Use Hose Lines.**   The fire code official is authorized to permit the removal of existing occupant-use hose lines where all of the following conditions exist:

1. Installation is not required by this code or the *International Building Code.*
2. The hose line will not be utilized by trained personnel or the fire department.
3. The remaining outlets are compatible with local fire department fittings.

CHANGE SIGNIFICANCE: IFC Section 905.3 identifies locations where standpipe systems are required. Both Class II and Class III standpipe systems consist of 1½-inch hose lines for occupant use. See the table below.

CLASSES OF STANDPIPE SYSTEMS

System	Hose and Connection Size
Class I	2½-inch hose connections to supply water for use by fire departments and those trained in handling heavy fire streams.
Class II	1½-inch hose stations to supply water for use primarily by the building occupants or by the fire department during initial response.
Class III	1½-inch hose stations to supply water for use by building occupants and 2½-inch hose connections for use by fire departments and those trained in handling heavy fire streams.

(From IFC Section 202)

In the past several editions of the code, there has been a reduction in the required installation of Class II systems. This is mainly a result of several key items:

- Maintenance of the hose lines frequently is marginal. The lack of maintenance has resulted in situations where an occupant is intending to fight a fire and grabs the hose nozzle; as the would-be firefighter approaches the fire, the occupant discovers that there is no water available because the cotton hose is mildewed and rotted. This puts the occupant in peril, because instead of exiting the building, the occupant is heading back toward the fire.
- OSHA requires that when hose lines for occupant use are provided, then the employees must be trained in their use. Training has been sporadic at best.
- Since most buildings are being equipped with a fire sprinkler system, the occupants should be instructed to let the fire sprinklers do their job while the occupants escape rather than stay near the fire to attempt to fight it.

Hose and nozzle removed from an existing Class II standpipe hose cabinet

Currently, Class II or III systems with 1½-inch hose lines are required in a few instances only. The 2015 IFC requires Class II systems with 1½-inch hose lines in

1. unsprinklered buildings with a story between 30 feet and 55 feet above the lowest level of fire department vehicle access (IFC 905.3.1, Exception 1), and
2. unsprinklered stages with an area greater than 1,000 square feet (IFC 905.3.4).

The requirement for installing 1½-inch hose lines is declining. There has been a shift in the philosophy of whether or not occupants should be asked to attempt to extinguish the fire or evacuate the structure. Therefore, removing the hose lines that would not be required if the building were built under today's code makes sense.

The new section states that the hose can be removed; it does not say that the system can be removed. The hose lines can be removed only if all specified criteria are met.

- The hose lines must not be required if the building were to be built under the current IFC and IBC.
- The facility does not have an on-site fire brigade or response team that consists of trained personnel who could utilize the hose lines.
- The fire department has determined that they will not utilize the hose lines.
- The connections that remain must be compatible with the fire department fittings. This will allow the fire department to connect its own fire hose and utilize these connections as a water source if it chooses to do so.

903.2.1

Fire Sprinklers in Group A Occupancies

CHANGE TYPE: Modification

CHANGE SUMMARY: When fire sprinklers are required in a Group A occupancy located on a story other than the level of exit discharge, fire sprinklers must be installed on all stories leading to all levels of exit discharge that are used by the Group A occupancy.

2015 CODE: 903.2.1 Group A. An automatic sprinkler system shall be provided throughout buildings and portions thereof used as Group A occupancies as provided in this section. For Group A-1, A-2, A-3 and A-4 occupancies, the automatic sprinkler system shall be provided throughout the ~~floor area~~ story where the fire area containing the Group A-1, A-2, A-3 or A-4 occupancy is located, and ~~in~~ throughout all ~~floors~~ stories from the Group A occupancy to, and including, the ~~nearest level~~ levels of exit

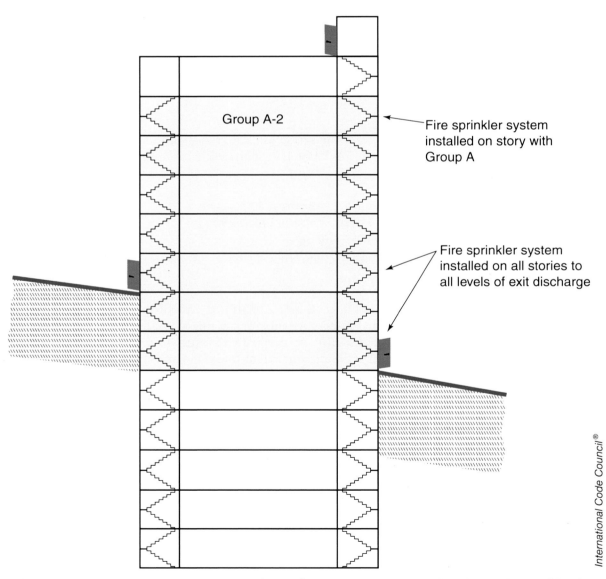

Group A-2

Fire sprinkler system installed on story with Group A

Fire sprinkler system installed on all stories to all levels of exit discharge

A fire sprinkler system is required in the story with the Group A occupancy and all stories leading to all levels of exit discharge

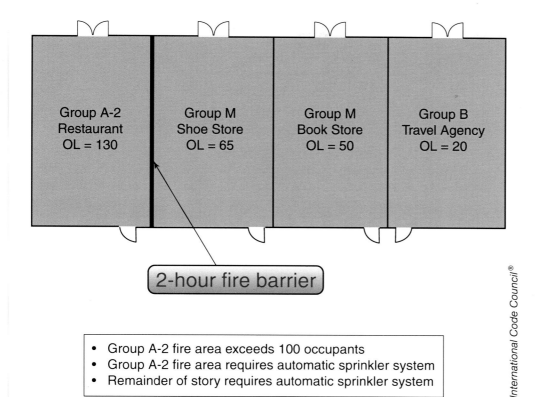

- Group A-2 fire area exceeds 100 occupants
- Group A-2 fire area requires automatic sprinkler system
- Remainder of story requires automatic sprinkler system

A fire sprinkler system is required in the Group A-2 fire area and throughout the entire story

discharge serving the Group A occupancy. For Group A-5 occupancies, the automatic sprinkler system shall be provided in spaces indicated in Section 903.2.1.5.

903.2.1.1 Group A-1. An automatic sprinkler system shall be provided for <u>fire areas containing</u> Group A-1 occupancies and <u>intervening floors of the building</u> where one of the following conditions exists:

1. The fire area exceeds 12,000 square feet (1115 m^2).
2. The fire area has an occupant load of 300 or more.
3. The fire area is located on a floor other than a level of exit discharge serving such occupancies.
4. The fire area contains a multitheater complex.

903.2.1.2 Group A-2. An automatic sprinkler system shall be provided for <u>fire areas containing</u> Group A-2 occupancies and <u>intervening floors of the building</u> where one of the following conditions exists:

1. The fire area exceeds 5,000 square feet (464 m^2).
2. The fire area has an occupant load of 100 or more.
3. The fire area is located on a floor other than a level of exit discharge serving such occupancies.

903.2.1 continues

903.2.1 continued

903.2.1.3 Group A-3. An automatic sprinkler system shall be provided for <u>fire areas containing</u> Group A-3 occupancies and <u>intervening floors of the building</u> where one of the following conditions exists:

1. The fire area exceeds 12,000 square feet (1115 m^2).
2. The fire area has an occupant load of 300 or more.
3. The fire area is located on a floor other than a level of exit discharge serving such occupancies.

903.2.1.4 Group A-4. An automatic sprinkler system shall be provided for <u>fire areas containing</u> Group A-4 occupancies and <u>intervening floors of the building</u> where one of the following conditions exists:

1. The fire area exceeds 12,000 square feet (1115 m^2).
2. The fire area has an occupant load of 300 or more.
3. The fire area is located on a floor other than a level of exit discharge serving such occupancies.

CHANGE SIGNIFICANCE: The requirement to provide fire sprinklers between the story that contains a Group A occupancy and the level of exit discharge now specifies that the fire sprinkler system must extend to all levels of exit discharge.

The 2012 IFC required the installation to the "nearest level" of exit discharge only. This would allow for the elimination of fire sprinklers to the next story even though one of the interior exit stairways did not exit the building at that story. This revision ensures that floors serving all exit discharges from the assembly occupancy are protected with fire sprinklers.

Also, revisions have been made to clarify the requirements in Sections 903.2.1.1 through 903.2.1.4. These changes specify that the fire sprinkler system is required in

- the entire story where there is a fire area containing the Group A occupancy, and
- all intervening stories to the levels of exit discharge.

For example, when the "fire area" is required to be sprinklered in Section 903.2.1.2, Section 903.2.1 requires the fire sprinkler system to extend beyond the fire area and include the entire floor of the building.

903.2.1.6
Assembly Occupancies on Roofs

CHANGE TYPE: Addition

CHANGE SUMMARY: Fire sprinklers are now required on all floors between the occupied roof and the level of exit discharge when assembly uses occur on the rooftop of buildings and the occupant load exceeds 100 for Group A-2 or 300 for other Group A occupancies.

2015 CODE: **903.2.1.6 Assembly Occupancies on Roofs.**   Where an occupied roof has an assembly occupancy with an occupant load exceeding 100 for Group A-2, and 300 for other Group A occupancies, all floors between the occupied roof and the *level of exit discharge* shall be equipped with an automatic sprinkler system in accordance with Section 903.3.1.1 or 903.3.1.2.

   Exception: Open parking garages of Type I or Type II construction.

CHANGE SIGNIFICANCE: Currently the code states that if a building has a fire area containing a Group A-1, A-2, A-3 or A-4 assembly on a floor other than the level of exit discharge, then the entire story and all stories to the level of exit discharge must be protected with a fire sprinkler system.

   Frequently, building owners will provide an open-air roof-top bar, lounge or other assembly-type use on the rooftop of a building. Since the rooftop of the building does not fit into the definition of a fire area, fire sprinklers are not provided. If that same use were inside the building, it would be protected with fire sprinklers along with all of the floor levels down to the levels of exit discharge.

   In order to provide a reasonable level of safety for the occupants on the roof, fire sprinklers are now required in the floor levels below the roof when the occupant loads exceed the normal threshold for fire sprinklers according to each particular occupancy. The fire sprinkler threshold for Group A-2 is an occupant load of 100; the threshold for Groups A-1, A-3 and A-4 is 300. The same thresholds are applied when the use is on the roof.

903.2.1.6 continues

Occupied roof with
assembly occupancy

Automatic sprinkler system required on
all floors between occupied roof and
level of exit discharge if
• roof is A-2 with occupant load
  exceeding 100 or
• occupant load exceeds
  300 for other assembly
  purposes.

Level of
exit discharge

International Code Council®

Assembly use on the roof protected by sprinklers on the stories beneath down to the level of exit discharge

903.2.1.6 continued   The rooftop itself is not required to be protected with a fire sprinkler system, but a fire sprinkler system is required on all stories down to the levels of exit discharge. The fire sprinkler system could be designed to either Section 903.3.1.1 or 903.3.1.2. This reference to Section 903.3.1.2 is included since the assembly use could occur on the rooftop of a residential building.

Open parking garages constructed of Type I or II construction are exempted from the fire sprinkler requirement. There is considerable data supporting the elimination of fire sprinklers in open parking garages. For example, two reports that evaluated fire behavior in parking garages are

1. 2006 NFPA Fire Data Report, "Structure and Vehicle Fires in General Vehicle Parking Garages" and
2. 2008 Parking Consultants Council Fire Safety Committee Report, "Parking Structure Fire Facts."

These reports provide the following conclusions:

- There is an average of only 660 fires per year in all types of parking garages in the U.S. This represents a mere 0.006 percent of all fires annually.
- There were no fire fatalities in open parking garages constructed of Construction Type I or II and an average of only 2 injuries per year.
- There was no structural damage in 98.7 percent of the fires in parking garages.
- Vehicle fires in parking garages typically do not spread from vehicle to vehicle. Fire spread from vehicle to vehicle occurred in only 7 percent of the incidents.

Fire sprinklers are required in occupancies other than open parking garages to protect the assembly occupancy above the fire and to protect the means of egress. Based on the inherent fire safety provided by open parking garages of Type I or II construction, a fire sprinkler system is not required when an assembly use is located on the roof.

CHANGE TYPE: Modification

CHANGE SUMMARY: When multiple Group A-1, A-2, A-3 or A-4 fire areas share egress paths, the occupant load will be combined for determining if a fire sprinkler system is required. The occupant load threshold is 300 or more.

2015 CODE: <u>**903.2.1.7 Multiple Fire Areas.** An automatic sprinkler system shall be provided where multiple fire areas of Group A-1, A-2, A-3 or A-4 occupancies share exit or exit access components and the combined occupant load of these fire areas is 300 or more.</u>

CHANGE SIGNIFICANCE: A single building can contain multiple Group A occupancies. It can often occur that each Group A occupancy is within its own fire area. The thresholds for fire sprinkler system installation are based on an evaluation of each fire area. However, when multiple Group A fire areas share the same egress paths, the fact that the occupants came from separate fire areas becomes less relevant.

903.2.1.7
Multiple Group A Fire Areas

903.2.1.7 continues

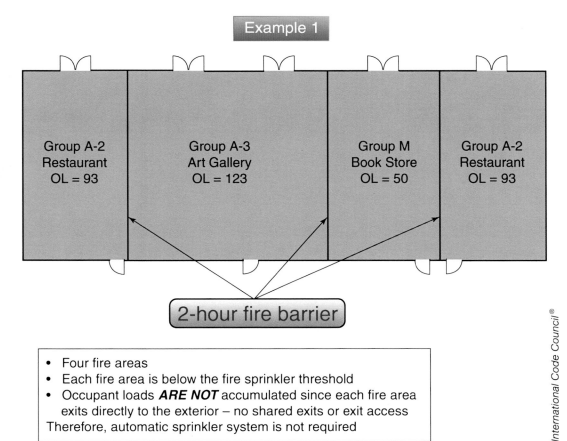

Example 1

| Group A-2 Restaurant OL = 93 | Group A-3 Art Gallery OL = 123 | Group M Book Store OL = 50 | Group A-2 Restaurant OL = 93 |

2-hour fire barrier

- Four fire areas
- Each fire area is below the fire sprinkler threshold
- Occupant loads **ARE NOT** accumulated since each fire area exits directly to the exterior – no shared exits or exit access
Therefore, automatic sprinkler system is not required

International Code Council®

The occupant load is combined only when the multiple Group A-1, A-2, A-3 or A-4 occupancies share exits or exit access

903.2.1.7 continued

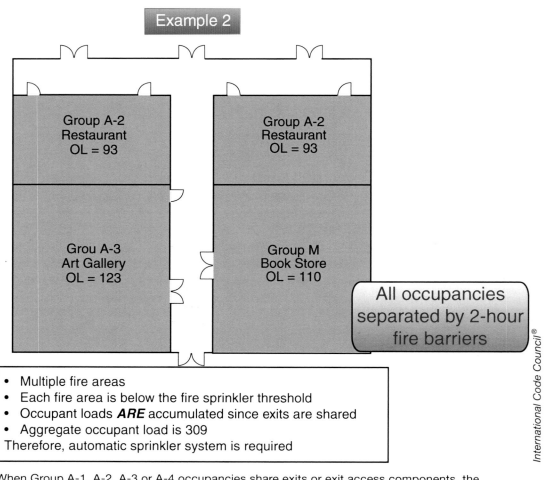

When Group A-1, A-2, A-3 or A-4 occupancies share exits or exit access components, the occupant loads are combined to determine fire sprinkler requirements

In Example 1, a single-story building has three Group A fire areas. Each fire area has an occupant load below the threshold that would require the installation of a fire sprinkler system. Each fire area is provided with an independent egress system.

In Example 2, a single-story building has three Group A fire areas. Again, each fire area has an occupant load below the threshold that would require the installation of a fire sprinkler system. However, in this case, all three fire areas share egress corridors and a lobby. When all three Group A fire areas evacuate at the same time, there are now 309 occupants in the egress system. This aggregate occupant load exceeds 300, so an automatic sprinkler system would be required.

This new section requires that the size of Group A fire areas is combined for Group A-1, A-2, A-3 and A-4 occupancies.

There are two criteria that must be met in order for the section to apply:

1. The multiple Group A-1, A-2, A-3 or A-4 fire areas must share exit or exit access components (see Example 1), and

2. The aggregate occupant load of the fire areas is 300 or more (see Example 2).

When the occupant loads of multiple fire areas are combined, the threshold for a fire sprinkler system is 300. This number applies whether the occupancies are Group A-3 or Group A-2. Even though a single Group A-2 fire area with an occupant load of 100 or more requires fire sprinklers, when Section 903.2.1.7 is utilized for multiple Group A-2 fire areas, the threshold occupant load is 300.

When this section is applied, only the occupant load is combined. As long as each fire area is below its individual threshold, the floor areas are not considered when the occupant load for the multiple Group A fire areas is being determined.

903.2.9

Commercial Motor Vehicles—Fire Sprinkler Requirements

CHANGE TYPE: Clarification

CHANGE SUMMARY: This code change provides a specific definition for commercial motor vehicles, which is applicable when the fire code official is determining whether a fire sprinkler system is required in specific occupancies.

2015 CODE: 903.2.9 Group S-1.   An automatic sprinkler system shall be provided throughout all buildings containing a Group S-1 occupancy where one of the following conditions exists:

1. A Group S-1 fire area exceeds 12,000 square feet (1115 m²).
2. A Group S-1 fire area is located more than three stories above grade plane.
3. The combined area of all Group S-1 fire areas on all floors, including any mezzanines, exceeds 24,000 square feet (2230 m²).
4. A Group S-1 fire area used for the storage of ~~commercial trucks or buses~~ commercial motor vehicles where the fire area exceeds 5,000 square feet (464 m²).
5. A Group S-1 occupancy used for the storage of upholstered furniture or mattresses exceeds 2,500 square feet (232 m²).

903.2.9.1 Repair Garages.   An automatic sprinkler system shall be provided throughout all buildings used as repair garages in accordance with Section 406.8 of the *International Building Code*, as shown:

1. Buildings having two or more stories above grade plane, including basements, with a fire area containing a repair garage exceeding 10,000 square feet (929 m²).
2. Buildings no more than one story above grade plane, with a fire area containing a repair garage exceeding 12,000 square feet (1115 m²).

Group S-1 repair garage for commercial motor vehicles *(Photo courtesy of Regional Transportation Commission of Southern Nevada, Las Vegas, NV)*

3. Buildings with repair garages servicing vehicles parked in basements.

4. A Group S-1 fire area used for the repair of ~~commercial trucks or buses~~ commercial motor vehicles where the fire area exceeds 5,000 square feet (464 m^2).

903.2.10.1 Commercial Parking Garages. An automatic sprinkler system shall be provided throughout buildings used for storage of ~~commercial trucks or buses~~ commercial motor vehicles where the fire area exceeds 5,000 square feet (464 m^2).

SECTION 202
GENERAL DEFINITIONS

Commercial Motor Vehicle. A motor vehicle used to transport passengers or property where the motor vehicle:

1. Has a gross vehicle weight rating of 10,000 pounds or more; or

2. Is designed to transport 16 or more passengers, including the driver.

CHANGE SIGNIFICANCE: The term "commercial trucks or buses" has been in the IFC since the 2000 edition. However, there has never been a definition to determine what constitutes commercial trucks and buses, which has caused some confusion. The application to buses was simple enough, but many states included pickup trucks, down to a 1-ton size, as commercial vehicles and required that they carry a commercial license plate.

To be considered a commercial motor vehicle, the IFC states the vehicle must be

- designed to carry 16 or more persons, or
- have a gross vehicle weight rating in excess of 10,000 pounds.

The definition places a passenger count on vehicles that will be considered "buses." The count of 15 passengers plus a driver will include many passenger transport shuttles in addition to the traditional bus such as Greyhound, Megabus, municipal buses and other similar buses.

The gross vehicle weight criteria correlates the large vehicle requirements in Section 1607.7 of the IBC and is consistent with DOT regulations 49 CFR 390.5.

The reason for the fire sprinkler system requirement is the increased fire load presented by these larger commercial vehicles. The quantity of fuel and the amounts of upholstered interior furnishings or plastics are greater in these vehicles. Large commercial vehicles may be carrying or transporting additional combustibles on-board, which also increases the fuel load.

The inclusion of the definition eliminates the guess work as to when the code official should apply the requirement for a fire sprinkler system.

903.2.11.3

Buildings 55 Feet or More in Height— Sprinklers Required

CHANGE TYPE: Clarification

CHANGE SUMMARY: This section has been revised to clarify how the height of a building is to be measured and that the section applies to buildings that have one or more stories. The exception for airport control towers has been deleted.

2015 CODE: 903.2.11.3 Buildings 55 Feet or More in Height.   An automatic sprinkler system shall be installed throughout buildings <u>that have one or more stories</u> with ~~a floor level having~~ an occupant load of 30 or more ~~that is~~ located 55 feet (16 764 mm) or more above the lowest level of fire department vehicle access, <u>measured to the finished floor</u>.

> **Exceptions:**
>> ~~1. Airport control towers.~~
>> ~~2.~~ <u>1.</u>   Open parking structures.
>> ~~3.~~ <u>2.</u>   Occupancies in Group F-2.

CHANGE SIGNIFICANCE: There are two primary revisions to this section. The first is a clarification of how height is to be measured. The terms "that have one or more stories" and "measured to the finished floor" have been added to this section. The use of the term "finished floor" clarifies that the 55-foot measurement is from the lowest level of fire department access to a finished floor containing 30 or more occupants. The following definitions for story and story above grade plane are from Chapter 2.

> **Story.**   That portion of a building included between the upper surface of a floor and the upper surface of the floor or roof next above (also see "Mezzanine" and Section 502.1 of the *International Building Code*). It is measured as the vertical distance from top to top of two successive tiers of beams or finished floor surfaces and, for the topmost story, from the top of the floor finish to the top of the ceiling joists or, where there is not a ceiling, to the top of the roof rafters.

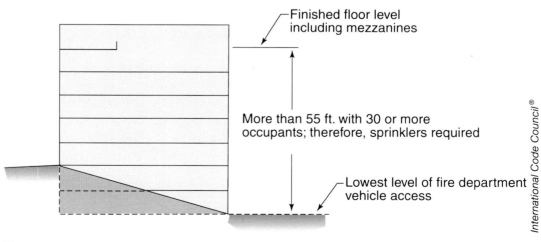

Finished floor level including mezzanines

More than 55 ft. with 30 or more occupants; therefore, sprinklers required

Lowest level of fire department vehicle access

International Code Council®

Measurement criteria

Story above Grade Plane.   Any story having its finished floor surface entirely above grade plane, or in which the finished surface of the floor next above is

1. more than 6 feet (1829 mm) above grade plane, or
2. more than 12 feet (3658 mm) above the finished ground level at any point.

The second change has been made to avoid conflict with the *International Building Code*. For the 2015 *International Fire Code*, the airport control tower exception to the sprinkler requirement in this section has been removed. This correlation is needed because the 2015 edition of the *International Building Code*, Section 412.3 now requires an automatic sprinkler system in all air traffic control towers with an occupiable floor 35 feet or more above the lowest level of fire department vehicle access.

903.3.1.1.1

Exempt Locations—Sprinklers Not Required when Automatic Fire Detection System Is Provided

Elevator control

CHANGE TYPE: Modification

CHANGE SUMMARY: This change introduces the concept of Machine Room-Less elevators (MRLs) to the IFC and provides correlation with ASME A17.1-2007/CSA B44-07. In the 2012 code, sprinkler exemptions are currently provided for elevator machine rooms and machinery spaces. This change expands the exemption to the control rooms and control spaces associated with occupant evacuation elevators. Additionally, the code has been changed regarding area smoke detection and fire command center requirements to reflect the defining of elevator control rooms and control spaces.

2015 CODE: 903.3.1.1 NFPA 13 Sprinkler Systems.   Where the provisions of this code require that a building or portion thereof be equipped throughout with an automatic sprinkler system in accordance with this section, sprinklers shall be installed throughout in accordance with NFPA 13 except as provided in Sections 903.3.1.1.1 and 903.3.1.1.2.

903.3.1.1.1 Exempt Locations.   Automatic sprinklers shall not be required in the following rooms or areas where such rooms or areas are protected with an approved automatic fire detection system in accordance with Section 907.2 that will respond to visible or invisible particles of combustion. Sprinklers shall not be omitted from any room merely because it is damp, of fire-resistance-rated construction or contains electrical equipment.

1. through 5. No significant changes
6. Machine rooms, and machinery spaces, <u>control rooms and control spaces</u> associated with occupant evacuation elevators designed in accordance with Section 3008 of the *International Building Code.*

907.2.13.1.1 Area Smoke Detection.   Area smoke detectors shall be provided in accordance with this section. Smoke detectors shall be connected to an automatic fire alarm system. The activation of any detector required by this section shall activate the emergency voice/alarm communication system in accordance with Section 907.5.2.2. In addition to smoke detectors required by Sections 907.2.1 through 907.2.10, smoke detectors shall be located as follows:

1. In each mechanical equipment, electrical, transformer, telephone equipment or similar room ~~which~~ <u>that</u> is not provided with sprinkler protection.
2. In each elevator machine room, <u>machinery space, control room and control space</u> and in elevator lobbies.

508.1.6 Required Features.   The fire command center shall comply with NFPA 72 and shall contain the following features:

1. through 12. No changes

13. An approved Building Information Card that contains, but is not limited to, the following information:

13.1. through 13.3. No changes

13.4. Exit access and exit stairway information that includes: number of exit accesses and exit stairways in building, each exit access and exit stairway designation and floors served, location where each exit access and exit stairway discharges, interior exit stairways that are pressurized, exit stairways provided with emergency lighting, each exit stairway that allows reentry, exit stairways providing roof access; elevator information that includes: number of elevator banks, elevator bank designation, elevator car numbers and respective floors that they serve, location of elevator machine rooms, <u>control rooms and control spaces,</u> location of sky lobby, location of freight elevator banks

13.5. through 18. No changes

CHANGE SIGNIFICANCE: This change will assist the code official in avoiding conflicts by correlating the requirements of the IFC with those of the ASME A17.1, *Safety Code for Elevators and Escalators.*

Within the elevator industry, the MRL design (Machine Room-Less elevators) has resulted in elevators, machines and controllers being located in rooms or spaces other than the traditional elevator machine rooms regulated by the IFC and IBC. In the 2012 edition of the code, controllers located in machine rooms are not required to be sprinklered. This change maintains the practice regarding controllers in machine rooms by not requiring sprinklers in control rooms or control spaces for MRLs used for occupant evacuation.

The ASME A17.1 underwent a substantial revision in 2005 to incorporate requirements for Machine Room-Less elevators. Included in the revisions were definitions to meet current technology and a refining of these definitions to provide the same protections or exemptions as the previous edition.

ASME A17.1 now includes definitions for elevator rooms and spaces that may contain various elevator apparatus. These definitions include

- A room outside the hoistway with an elevator machine is a "machine room";
- A room or space outside the hoistway with a motor controller and not a machine is a "control room" or "control space";
- A space where a machine and motor controller are located inside the hoistway, the hoistway is a "machinery space."

In addition, within the elevator industry, the following designs are widely used:

- Machinery and control spaces may have doors;
- Elevator controllers include the operation controller and motion controller, which may be separated from the location of the elevator machine and be located in separate elevator rooms and spaces;

903.3.1.1.1 continues

903.3.1.1.1 continued

- Machine rooms and control rooms are full-body spaces with doors that may have room sprinklers and fire detection apparatus; control and machinery spaces typically would not;

- Machine rooms and control rooms typically require room ventilation and cooling; machinery and control spaces typically do not;

- Machinery spaces inside the hoistway are covered by the code's hoistway requirements;

- Elevator machines and electrical apparatus in spaces other than the hoistway or rooms may require standby power for apparatus cooling equipment.

CHANGE TYPE: Addition

CHANGE SUMMARY: This new section provides criteria for not installing sprinklers in bathrooms of specific Group R occupancies.

2015 CODE: **903.3.1.1.2 Bathrooms.**   In Group R occupancies, other than Group R-4 occupancies, sprinklers shall not be required in bathrooms that do not exceed 55 square feet (5 m^2) in area and are located within individual dwelling units or sleeping units, provided that walls and ceilings, including the walls and ceilings behind a shower enclosure or tub, are of noncombustible or limited-combustible materials with a 15-minute thermal barrier rating.

CHANGE SIGNIFICANCE: Bathrooms have been exempted from the requirements of NFPA 13, Standard for the Installation of Sprinkler Systems from 1976 until the 2013 edition. However, the recently published NFPA 13-2013 (Section 8.15.8.1.1) allows the omission of sprinklers in bathrooms in hotels and motels only, not apartments. To maintain consistency with current practice, this section has been added to the IFC, reinstating the exception for sprinklers when bathrooms in Group R Division 1, 2 and 3 occupancies conform to all of the following:

- Do not exceed 55 square feet in area;
- Are located within individual dwelling units or sleeping units;
- Have walls and ceilings, including walls and ceilings behind any shower enclosure or tub, of noncombustible or limited-combustible materials with a 15-minute thermal barrier rating.

903.3.1.1.2
Bathrooms Exempt from Sprinkler Requirements

Bathroom less than 55 square feet

903.3.1.2

NFPA 13R Sprinkler Systems

CHANGE TYPE: Clarification

CHANGE SUMMARY: This change correlates Group R limitations on height with the scope of NFPA 13R.

2015 CODE: 903.3.1.2 NFPA 13R Sprinkler Systems. Automatic sprinkler systems in Group R occupancies up to and including four stories in height ~~above grade plane~~ in buildings not exceeding 60 feet (18 288 mm) in height above grade plane shall be permitted to be installed throughout in accordance with NFPA 13R.

The number of stories of Group R occupancies constructed in accordance with Sections 510.2 and 510.4 of the *International Building Code* shall be measured from the horizontal assembly creating separate buildings.

CHANGE SIGNIFICANCE: The change clarifies that the limits for using NFPA 13R, Standard for the Installation of Sprinkler Systems in Low-Rise Residential Occupancies sprinkler systems are

- maximum of four stories in height, and
- maximum of 60 feet in height above grade plane.

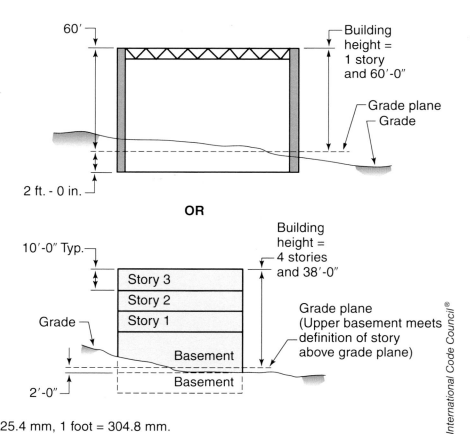

For SI: 1 inch = 25.4 mm, 1 foot = 304.8 mm.

Limits for using 13R sprinkler systems

International Code Council®

Additionally, today's trend in construction is the designing of podium buildings with horizontal assemblies between the lower levels and the residential floors. The new second paragraph addresses the scenario where NFPA 13R systems are desired to be installed in residential buildings using the podium building provisions in Sections 510.2 (Horizontal building separation allowance) and 510.4 (Parking beneath Group R) of the *International Building Code.*

From a fire safety standpoint, the horizontal separation is significant enough to justify the measurement of number of stories from that point. This allowance is related to the number of stories only, not the height in feet above grade plane.

The following definition of story above grade plane is from Chapter 2 of the IFC.

> **Story above Grade Plane.** Any story having its finished floor surface entirely above grade plane, or in which the finished surface of the floor next above is
>
> 1. more than 6 feet (1829 mm) above grade plane, or
> 2. more than 12 feet (3658 mm) above the finished ground level at any point.

903.3.1.2.2, 1027.6, 1104.22

NFPA 13R Sprinkler Systems—Open-Ended Corridors (Breezeways)

Sprinklers protecting an open-ended corridor

CHANGE TYPE: Addition

CHANGE SUMMARY: The intent of Section 903.3.1.2.2 is to clarify that when an NFPA 13R sprinkler system is used, additional heads are required in the open-ended corridor (breezeway). An associated change is in Section 1027.6 for exterior stairways and ramps. To correlate the open-ended corridor concept in existing buildings with these changes, Section 1104.22 has been changed through the deletion of exception 1, which allows the open-ended corridor criteria to dictate the solution.

2015 CODE:

SECTION 202
GENERAL DEFINITIONS

Open-Ended Corridor. <u>An interior corridor that is open on each end, and connects to an exterior stairway or ramp at each end with no intervening doors or separation from the corridor.</u>

903.3.1.2.2 Open-Ended Corridors. <u>Sprinkler protection shall be provided in open-ended corridors and associated exterior stairways and ramps as specified in Section 1027.6, Exception 3.</u>

~~1026.6~~<u>1027.6</u> **Exterior Stairway and Ramp Protection.** Exterior exit stairways and ramps shall be separated from the interior of the building as required in Section ~~1022.7~~<u>1023.2</u>. Openings shall be limited to those necessary for egress from normally occupied spaces. <u>Where a vertical plane projecting from the edge of an exterior stairway or ramp and landings is exposed by other parts of the building at an angle of less than 180 degrees (3.14 rad), the exterior wall shall be rated in accordance with Section 1023.7.</u>

Exceptions:

1. Separation from the interior of the building is not required for occupancies, other than those in Group R-1 or R-2, in buildings that are no more than two stories above grade plane where a level of exit discharge serving such occupancies is the first story above grade plane.

2. Separation from the interior of the building is not required where the exterior exit stairway or ramp is served by an exterior ramp or balcony that connects two remote exterior stairways or other approved exits, with a perimeter that is not less than 50 percent open. To be considered open, the opening shall be a minimum of 50 percent of the height of the enclosing wall, with the top of the openings no less than 7 feet (2134 mm) above the top of the balcony.

3. Separation from the open-ended corridor of the building is not required for exterior exit stairways or ramps, provided that Items 3.1 through 3.5 are met:

   3.1. The building, including <u>open-ended</u> corridors, and stairways ~~or~~ <u>and</u> ramps, shall be equipped throughout with an automatic sprinkler system in accordance with Section 903.3.1.1 or 903.3.1.2.

   3.2. The open-ended corridors comply with Section 1020.

3.3. The open-ended corridors are connected on each end to an exterior exit stairway or ramp complying with Section 1027.

3.4. The exterior walls and openings adjacent to the exterior exit stairway or ramp comply with Section 1023.7.

3.5. At any location in an open-ended corridor where a change of direction exceeding 45 degrees (0.79 rad) occurs, a clear opening of not less than 35 square feet (3.3 m^2) or an exterior stairway or ramp shall be provided. Where clear openings are provided, they shall be located so as to minimize the accumulation of smoke or toxic gases.

~~1104.21~~ **1104.22 Exterior Stairway Protection.** Exterior exit stairways shall be separated from the interior of the building as required in Section 1027.6. Openings shall be limited to those necessary for egress from normally occupied spaces.

Exceptions:

1. Separation from the interior of the building is not required for buildings that are two stories or less above grade where the level of exit discharge serving such occupancies is the first story above grade.

2. Separation from the interior of the building is not required where the exterior stairway is served by an exterior balcony that connects two remote exterior stairways or other approved exits, with a perimeter that is not less than 50 percent open. To be considered open, the opening shall be not less than 50 percent of the height of the enclosing wall, with the top of the opening not less than 7 feet (2134 mm) above the top of the balcony.

3. Separation from the interior of the building is not required for an exterior stairway located in a building or structure that is permitted to have unenclosed interior stairways in accordance with Section 1023.

4. Separation from the interior of the building is not required for exterior stairways connected to open-ended *corridors*, provided that:

    4.1. ~~The building, including corridors and stairways, is equipped throughout with an automatic sprinkler system in accordance with Section 903.3.1.1 or 903.3.1.2.~~

    ~~4.2.~~ **4.1.** The open-ended corridors comply with Section 1020.

    ~~4.3.~~ **4.2.** The open-ended corridors are connected on each end to an exterior *exit* stairway complying with Section 1027.

    ~~4.4.~~ **4.3.** At any location in an open-ended corridor where a change of direction exceeding 45 degrees (0.79 rad) occurs, a clear opening of not less than 35 square feet (3 m^2) or an exterior stairway shall be provided. Where clear openings are provided, they shall be located so as to minimize the accumulation of smoke or toxic gases.

903.3.1.2.2, 1027.6, 1104.22 continues

903.3.1.2.2, 1027.6, 1104.22
continued

CHANGE SIGNIFICANCE: This section adds language clarifying that open-ended corridors (breezeways) in buildings equipped with NFPA 13R, Standard for the Installation of Sprinkler Systems in Low-Rise Residential Occupancies sprinkler systems need to be protected. The designer's option to create an open-ended corridor (breezeway) that eliminates the requirement for walls, doors, and opening protection at the ends to separate this exit access component from an exterior stair does not preclude the need to extend the sprinkler protection from the living space into the open-ended corridor.

The addition of this language to the IFC creates a requirement on NFPA 13R sprinkler systems that is greater than the standard requires. Accordingly, the plans examiner needs to be alert when reviewing sprinklered buildings with breezeways so that this requirement is caught during plan check and not in the field. In cold climates, the protection of this breezeway will require special attention to the design and type of sprinkler head to be used if an anti-freeze loop or dry system is to be avoided.

With this set of changes the IFC has been correlated by the following:

• A new definition in Chapter 2 for "Open-Ended Corridor."

• A specification that sprinkler protection shall be provided in open-ended corridors; see Section 903.3.1.2.2.

• Revisions to Section 1027 that add clarity regarding separation requirements for open-ended corridors (breezeways).

• A deletion of the sprinkler requirement in Chapter 11, Section 1104.22 that previously required an existing building with open-ended corridors (breezeways) to be retroactively sprinklered.

903.3.8
Limited Area Sprinkler Systems

CHANGE TYPE: Modification

CHANGE SUMMARY: This change reduces the number of sprinkler heads that can be used in a limited area sprinkler system from 20 heads to 6 heads. This change provides additional criteria regarding the use of these systems.

2015 CODE: ~~903.3.5.1.1 Limited Area Sprinkler Systems. Limited area sprinkler systems serving fewer than 20 sprinklers on any single connection are permitted to be connected to the domestic service where a wet automatic standpipe is not available. Limited area sprinkler systems connected to domestic water supplies shall comply with each of the following requirements:~~

~~1. Valves shall not be installed between the domestic water riser control valve and the sprinklers.~~

   ~~**Exception:** An approved indicating control valve supervised in the open position in accordance with Section 903.4.~~

~~2. The domestic service shall be capable of supplying the simultaneous domestic demand and the sprinkler demand required to be hydraulically calculated by NFPA 13NFPA 13D NFPA 13R~~

903.3.8 Limited Area Sprinkler Systems. Limited area sprinkler systems shall be in accordance with the standards listed in Section 903.3.1 except as provided in Sections 903.3.8.1 through 903.3.8.5.

903.3.8.1 Number of Sprinklers. Limited area sprinkler systems shall not exceed six sprinklers in any single fire area.

903.3.8.2 Occupancy Hazard Classification. Only areas classified by NFPA 13 as Light Hazard or Ordinary Hazard Group 1 shall be permitted to be protected by limited area sprinkler systems.

903.3.8.3 Piping Arrangement. Where a limited area sprinkler system is installed in a building with an automatic wet standpipe system, sprinklers shall be supplied by the standpipe system. Where a limited area sprinkler system is installed in a building without an automatic wet standpipe system, water shall be permitted to be supplied by the plumbing system provided that the plumbing system is capable of simultaneously supplying domestic and sprinkler demands.

903.3.8.4 Supervision. Control valves shall not be installed between the water supply and sprinklers unless the valves are of an approved indicating type that are supervised or secured in the open position.

903.3.8.5 Calculations. Hydraulic calculations in accordance with NFPA 13 shall be provided to demonstrate that the available water flow and pressure are adequate to supply all sprinklers installed in any single fire area with discharge densities corresponding to the hazard classification.

Limited area sprinkler protection

International Code Council®

903.3.8 continues

903.3.8 continued

CHANGE SIGNIFICANCE: This section has been revised and criteria added to more clearly define the intent of where limited area sprinkler systems may be used. Limited area sprinkler systems are special systems that may be supplied from a building's domestic plumbing system. Examples of the use of limited area sprinkler systems in buildings not equipped with automatic fire sprinklers are

- the sprinklering of a walk-in cooler box at a convenience store as required by IBC Section 2603.4.1.2,
- the storage of medical gases at dentists' offices using interior storage in accordance with IFC Section 5306.2.2 and
- the use of IBC Table 509 for incidental uses

This change is intended to eliminate the potential for multiple limited area sprinkler systems within a single fire area. The IFC definition of Fire Area:

> **[B] Fire Area.** The aggregate floor area enclosed and bounded by *fire walls, fire barriers, exterior walls,* or *horizontal assemblies* of a building. Areas of the building not provided with surrounding walls shall be included in the fire area if such areas are included within the horizontal projection of the roof or floor next above.

The criteria for using limited area sprinkler systems include

- The number of sprinkler heads shall not exceed 6 within a fire area.
- Only areas classified by NFPA 13 as Light Hazard or Ordinary Hazard Group 1 are permitted to be protected by a limited area sprinkler system.
- Water to the sprinkler heads shall be provided from an automatic wet standpipe system if one is present in the building. Alternatively, the water may be supplied by the domestic plumbing system providing the domestic system is capable of simultaneously supplying both the domestic and sprinkler demands.
- No control valves are to be installed unless they are of an approved indicating type that are supervised or secured in the open position.
- Hydraulic calculations shall be provided that prove the piping arrangement and water supply are adequate to meet the sprinkler demand for all sprinklers installed within the fire area with discharge densities corresponding to the hazard classification.

904.2, 904.11
Automatic Water Mist Systems

CHANGE TYPE: Addition

CHANGE SUMMARY: This change recognizes automatic water mist systems as an alternative, on a limited basis, to automatic fire sprinkler systems. Automatic water mist systems are most commonly used for special protection applications for special hazard applications such as computer room subfloors and machinery spaces.

2015 CODE: 904.2 ~~Where Required~~ Permitted. Automatic fire-extinguishing systems installed as an alternative to the required automatic sprinkler systems of Section 903 shall be approved by the fire code official. ~~Automatic fire-extinguishing systems shall not be considered alternatives for the purposes of exceptions or reductions allowed by other requirements of this code.~~

904.2.1 Restriction on Using Automatic Sprinkler System Exceptions or Reductions. Automatic fire-extinguishing systems shall not be considered alternatives for the purposes of exceptions or reductions allowed for automatic sprinkler systems or by other requirements of this code.

904.11 Automatic Water Mist Systems. Automatic water mist systems shall be permitted in applications that are consistent with the applicable listing or approvals and shall comply with Sections 904.11.1 through 904.11.3.

904.11.1 Design and Installation Requirements. Automatic water mist systems shall be designed and installed in accordance with Sections 904.11.1.1 through 904.11.1.4.

904.11.1.1 General. Automatic water mist systems shall be designed and installed in accordance with NFPA 750 and the manufacturer's instructions.

904.11.1.2 Actuation. Automatic water mist systems shall be automatically actuated.

904.11.1.3 Water Supply Protection. Connections to a potable water supply shall be protected against backflow in accordance with the *International Plumbing Code.*

904.11.1.4 Secondary Water Supply. Where a secondary water supply is required for an automatic sprinkler system, an automatic water mist system shall be provided with an approved secondary water supply.

904.11.2 Water Mist System Supervision and Alarms. Supervision and alarms shall be provided as required for automatic sprinkler systems in accordance with Section 903.4.

904.11.2.1 Monitoring. Monitoring shall be provided as required for automatic sprinkler systems in accordance with Section 903.4.1.

904.2, 904.11 continues

Water mist nozzle discharge   *(Photo courtesy of Tyco Fire Protection Products)*

904.2, 904.11 continued

904.11.2.2 Alarms. Alarms shall be provided as required for automatic sprinkler systems in accordance with Section 903.4.2.

904.11.2.3 Floor Control Valves. Floor control valves shall be provided as required for automatic sprinkler systems in accordance with Section 903.4.3.

904.11.3 Testing and Maintenance. Automatic water mist systems shall be tested and maintained in accordance with Section 901.6.

SECTION 202
GENERAL DEFINITIONS

Automatic Water Mist System. A system consisting of a water supply, a pressure source, and a distribution piping system with attached nozzles, which, at or above a minimum operating pressure, defined by its listing, discharges water in fine droplets meeting the requirements of NFPA 750 for the purpose of the control, suppression or extinguishment of a fire. Such systems include wet-pipe, drypipe and pre-action types. The systems are designed as engineered, preengineered, local-application or total flooding systems.

CHANGE SIGNIFICANCE: A water mist system is a fire-protection system that uses very small water droplets (i.e., water mist). The small water droplets allow the water mist to control, suppress or extinguish fires. When approved by the fire code official, Section 904.2 recognizes water mist as an alternative, in some applications, to automatic fire sprinkler systems. However, no exceptions, reductions or "trade-offs" for water mist systems are granted or permitted by Sections 904.2.1, as automatic water mist systems are not considered equivalent to automatic sprinkler systems.

Section 904.11 states: "Automatic water mist systems shall be permitted in applications that are consistent with sections 904.11.1 through 904.11.3." The first subsection 904.11.1 requires compliance with NFPA 750, Standard on Water Mist Fire Protection Systems. Chapter 5 of NFPA 750 requires that the design of a water mist system must be based on results of comprehensive fire tests, conducted by an internationally recognized laboratory. This performance-based approach obligates manufacturers to test and demonstrate their systems performances. The tests need to confirm that compartment size, ventilation, obstructions, fuel quantities, and other features of the application match the assumptions in the test protocol.

In addition to the requirements of NFPA 750 this new section requires:

- Automatic activation;
- Protection of the potable water supply from backflow in accordance with the *International Plumbing Code*;
- When a secondary water supply is required for the building, the automatic water mist system shall also be provided with an approved secondary water supply;
- Monitoring in accordance with Section 903.4.1 (Monitoring);
- Alarms in accordance with Section 903.4.2 (Alarms);
- Floor control valves in accordance with Section 903.4.3 (Floor control valves);
- Testing and maintenance in accordance with Section 901.6 (Inspection, Testing and Maintenance).

CHANGE TYPE: Addition

CHANGE SUMMARY: UL 300A has been added to the IFC, and the new definition of Institutional Occupancy Group 2 Condition 1 (Nursing Homes, Assisted Living, etc.) from the IBC for an extinguishing system within the domestic cooking hood of such occupancy has been incorporated.

2015 CODE: <u>**904.13 Domestic Cooking Systems in Group I-2 Condition 1.**   In Group I-2 Condition 1 occupancies where cooking facilities are installed in accordance with Section 407.2.6 of the *International Building Code*, the domestic cooking hood provided over the cooktop or range shall be equipped with an automatic fire-extinguishing system of a type recognized for protection of domestic cooking equipment. Preengineered automatic extinguishing systems shall be tested in accordance with UL 300A and listed and labeled for the intended application. The system shall be installed in accordance with this code, its listing and the manufacturer's instructions.</u>

<u>**904.13.1 Manual System Operation and Interconnection.**   Manual actuation and system interconnection for the hood suppression system shall be installed in accordance with Sections 904.12.1 and 904.12.2.</u>

<u>**904.13.2 Portable Fire Extinguishers for Domestic Cooking Equipment in Group I-2 Condition 1.**   A portable fire extinguisher complying with Section 906 shall be installed within 30-foot (9144 mm) distance of travel from domestic cooking appliances.</u>

CHANGE SIGNIFICANCE: The IBC has now incorporated criteria for common area cooking in Group I-2 Condition 1 Occupancies. These criteria include:

- Reference to UL 300A, Outline of Investigation for Extinguishing System Units for Residential Range Top Cooking Surfaces;
- Manual operations and interconnections in accordance with Sections 904.12.1 and 904.12.2;
- A portable fire extinguisher in accordance with Section 906 within 30 feet travel distance of the domestic cooking appliance.

As nursing homes move away from institutional models, it is critical that they have a functioning kitchen that can serve as the focal point. Instead of a large, centralized, institutional kitchen where all meals are prepared and delivered to a central dining room or a resident's room, the new "household model" nursing home uses de-centralized kitchens and small dining areas to create the feeling and focus of home. It is particularly important for people with dementia to have spaces that look familiar, like the kitchen in their former home, to increase their understanding and ability to function at their highest level.

Allowing kitchens that serve a small, defined group of residents to be open to common spaces is critically important to enhancing the feeling and memories of home for older adults. This allows residents to see and smell the food being prepared, which can enhance their appetites and

904.13 continues

904.13
Domestic Cooking Systems in Group I-2 Condition 1 Occupancies

Domestic cooking hood with automatic extinguishing system *(Photograph courtesy of Denlar Fire Protection)*

904.13 continued evoke positive memories. Some residents, based on their abilities and cognition level, may even be able to participate in food preparation activities such as stirring, measuring ingredients, peeling vegetables, or folding towels. This becomes a social activity, where they can easily converse with the staff member cooking, as well as a way for the resident to maintain her or his functional abilities and to feel that he or she is still an important contributing member of society.

The revised definition for this Occupancy Classification reads as follows:

Institutional Group I-2. Institutional Group I-2 occupancy shall include buildings and structures used for medical care on a 24-hour basis for more than five persons who are not capable of self-preservation. This group shall include, but not be limited to, the following:

Foster care facilities
Detoxification facilities
Hospitals
Nursing homes
Psychiatric hospitals

Occupancy Conditions. Buildings of Group I-2 shall be classified as one of the following occupancy conditions:

Condition 1. This occupancy condition shall include facilities that provide nursing and medical care but do not provide emergency care, surgery, obstetrics, or in-patient stabilization units for psychiatric or detoxification, including, but not limited to nursing homes and foster care facilities.

Condition 2. This occupancy condition shall include facilities that provide nursing and medical care and could provide emergency care, surgery, obstetrics, or inpatient stabilization units for psychiatric or detoxification, including, but not limited to hospitals.

Five or Fewer Persons Receiving Medical Care. A facility with five or fewer persons receiving medical care shall be classified as Group R-3 or shall comply with the *International Residential Code* provided an *automatic sprinkler system* is installed in accordance with Section 903.3.1.3 or with Section P2904 of the *International Residential Code.*

IBC Section 407.2.6 states:

407.2.6 Nursing Home Cooking Facilities. In Group I-2, Condition 1, occupancies, rooms or spaces that contain a cooking facility with domestic cooking appliances shall be permitted to be open to the corridor where all of the following criteria are met:

1. The number of care recipients housed in the smoke compartment is not greater than 30.
2. The number of care recipients served by the cooking facility is not greater than 30.

3. Only one cooking facility area is permitted in a smoke compartment.

4. The types of domestic cooking appliances permitted are limited to ovens, cooktops, ranges, warmers and microwaves.

5. The corridor is a clearly identified space delineated by construction or floor pattern, material or color.

6. The space containing the domestic cooking facility shall be arranged so as not to obstruct access to the required exit.

7. A domestic cooking hood installed and constructed in accordance with Section 505 of the *International Mechanical Code* is provided over the cooktop or range.

8. The domestic cooking hood provided over the cooktop or range shall be equipped with an automatic fire-extinguishing system of a type recognized for protection of domestic cooking equipment. Preengineered automatic extinguishing systems shall be tested in accordance with UL 300A and listed and labeled for the intended application. The system shall be installed in accordance with this code, its listing and the manufacturer's instructions.

9. A manual actuation device for the hood suppression system shall be installed in accordance with Sections 904.12.1 and 904.12.2.

10. An interlock device shall be provided such that upon activation of the hood suppression system, the power or fuel supply to the cooktop or range will be turned off.

11. A shut-off for the fuel and electrical power supply to the cooking equipment shall be provided in a location that is accessible only to staff.

12. A timer shall be provided that automatically deactivates the cooking appliances within a period of not more than 120 minutes.

13. A portable fire extinguisher shall be installed in accordance with Section 906 of the *International Fire Code.*

907.1.2

Fire Alarm Shop Drawings—Design Minimum Audibility Level

CHANGE TYPE: Addition

CHANGE SUMMARY: The fire alarm designer is now required to provide the design minimum audibility level for occupant notification, and the phrase "where applicable" has been added to the charging statement to clarify that not all items shown in the list may be applicable for every installation.

2015 CODE: 907.1.2 Fire Alarm Shop Drawings. Shop drawings for fire alarm systems shall be submitted for review and approval prior to system installation, and shall include, but not be limited to, all of the following <u>where applicable to the system being installed</u>:

1. A floor plan that indicates the use of all rooms.
2. Locations of alarm-initiating devices.
3. Locations of alarm notification appliances, including candela ratings for visible alarm notification appliances.
4. <u>Design minimum audibility level for occupant notification.</u>
5. Location of fire alarm control unit, transponders and notification power supplies.
6. Annunciators.
7. Power connection.
8. Battery calculations.
9. Conductor type and sizes.
10. Voltage drop calculations.
11. Manufacturers' data sheets indicating model numbers and listing information for equipment, devices and materials.

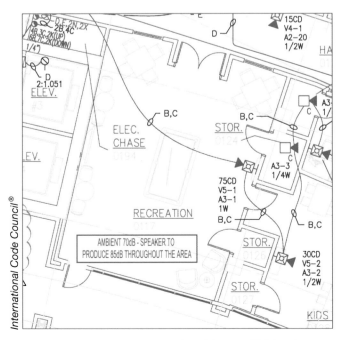

Recreation area with designer's minimum dBA level

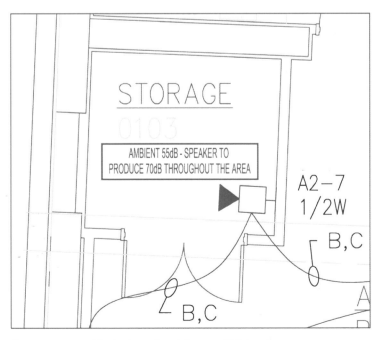

Storage area with designer's minimum dBA level

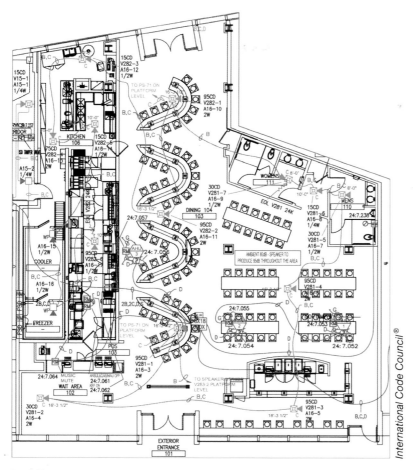

Restaurant area with designer's minimum dBA level

12. Details of ceiling height and construction.

13. The interface of fire safety control functions.

14. Classification of the supervising station.

CHANGE SIGNIFICANCE: The "where applicable" addition is necessary to clarify that only those items applicable to the system being installed are required to be submitted. For example, if the system is only to monitor a sprinkler system and no fire alarm notification appliances are required, there is no need to provide voltage-drop calculations or minimum audibility levels that the system will be designed to meet. Alternatively, a small remodel may not require the inclusion of all the items listed due to the limited nature of the work.

The addition of a requirement for the designer to designate the design minimum audibility level for occupant notification is intended to assist the plans examiner and field inspector as well as the fire alarm installer. The proponent of this code change intended for the designer to designate what the minimum audibility level is for each different and distinct area being protected by the fire alarm system. By stating the decibel level needed, the designer will give the plans examiner and field inspector information that is not commonly provided. This will enable the examiner

907.1.2 continues

907.1.2 continued and inspector to make decisions based on a given decibel rather than the performance criteria of NFPA 72, *National Fire Alarm and Signaling Code* for either a public mode signal or a private mode signal. Typically, the fire alarm designer makes several decisions that may or may not be reflected on the plans: first, which areas are public modes and which are private modes; second, the decibel level that is needed to meet code for each different and distinct area. Unfortunately, this analysis is not normally provided by the system designer and therefore the plans examiner and field inspector are left to form their own decisions as to what is the correct amount of audibility. This change will provide more information to facilitate fire alarm approvals, especially at final inspection when audibility tests are conducted. The proponent has stated that it is not intended to simply have the designer state that the audibility levels must meet the audibility requirements of NFPA 72, but to list specific minimum sound levels the system must meet.

NFPA 72 Section 3.3.185.1 defines Private Operating Mode as "Audible or visual signaling only to those persons directly concerned with the implementation and direction of emergency action initiation and procedure in the area protected by the fire alarm system (SIG-NAS)."

NFPA 72 Section 3.3.185.2 defines Public Operating Mode as "Audible or visual signaling to occupants or inhabitants of the area protected by the fire alarm system (SIG-NAS)."

NFPA 72 Section 18.4.3.1 states:

To ensure that audible public mode signals are clearly heard, unless otherwise permitted by 18.4.3.2 through 18.4.3.5, they shall have a sound level at least 15dB above the average ambient sound level or 5 dB above the maximum sound level having a duration of at least 60 seconds, whichever is greater, measured 5 ft (1.5 m) above the floor in the area required to be served by the system using the A-weighted scale (dBA).

Alternatively, NFPA 72 Section 18.4.4.1 states:

To ensure that audible private mode signals are clearly heard they shall have a sound level at least 10dB above the average ambient sound level or 5 dB above the maximum sound level having a duration of at least 60 seconds, whichever is greater, measured 5 ft (1.5 m) above the floor in the area required to be served by the system using the A-weighted scale (dBA).

CHANGE TYPE: Modification

CHANGE SUMMARY: The threshold for requiring a manual fire alarm system has been raised from 30 occupants to 50. The emergency voice/alarm communication system requirement has been raised to 100 occupants.

2015 CODE: 907.2.3 Group E.   A manual fire alarm system that initiates the occupant notification signal utilizing an emergency voice/alarm communication system meeting the requirements of Section 907.5.2.2 and installed in accordance with Section 907.6 shall be installed in Group E occupancies. When automatic sprinkler systems or smoke detectors are installed, such systems or detectors shall be connected to the building fire alarm system.

Exceptions:

1. A manual fire alarm system is not required in Group E occupancies with an occupant load of ~~30~~ <u>50</u> or less.

2. <u>Emergency voice/alarm communication systems meeting the requirements of Section 907.5.2.2 and installed in accordance with Section 907.6 shall not be required in Group E occupancies with occupant loads of 100 or less, provided that activation of the manual fire alarm system initiates an approved occupant notification signal in accordance with Section 907.5.</u>

3. Manual fire alarm boxes are not required in Group E occupancies where all of the following apply:

    3.1. Interior corridors are protected by smoke detectors.

    3.2. Auditoriums, cafeterias, gymnasiums, and similar areas are protected by heat detectors or other approved detection devices.

    3.3. Shops and laboratories involving dusts or vapors are protected by heat detectors or other approved detection devices.

4. Manual fire alarm boxes shall not be required in Group E occupancies where all of the following apply:

    4.1. The building is equipped throughout with an approved automatic sprinkler system installed in accordance with Section 903.3.1.1.

    4.2. The emergency voice/alarm communication system will activate on sprinkler water flow.

    4.3. Manual activation is provided from a normally occupied location.

CHANGE SIGNIFICANCE: Many small schools or day-care facilities consist of one or two rooms. For such small buildings, there is no need to install a notification system to warn occupants of fires or other emergencies, as occupants are typically in close visual or audible contact with all occupied spaces and with each other. This arrangement provides for adequate means to notify all occupants of the building of potential hazardous

907.2.3 continues

907.2.3
Group E Manual Fire Alarm System

Manual fire alarm—double-action pull station

International Code Council®

907.2.3 continued   conditions to initiate emergency actions, including evacuation. The use of an occupant load of 50 to trigger the requirement for a manual fire alarm system is now correlated to the same quantity of occupants used to require a second exit in group E occupancies.

Exception 2 has been added to set the threshold for emergency voice/alarm communication systems at a higher threshold than the 50 occupants or less that is used to require a manual fire alarm system. The threshold of 100 or less occupants has been selected to take into account educational occupancies containing multiple floor, fire areas, and egress paths. The occupants of these larger, more complex buildings may need to be provided with detailed or custom instructions on alternative courses of action other than those practiced in the standard evacuation drills.

In summary, the changes made to Section 907.2.3 result in the following for Group E occupancies:

When the occupant load is 50 or less, a manual fire alarm system is not required.

When the occupant load is 51 to 100, a manual fire alarm system is required.

When the occupant load is 101 or more, a manual fire alarm system with emergency voice/alarm communication system is required.

CHANGE TYPE: Clarification

CHANGE SUMMARY: The change to 907.2.6 Exception 2 links the use of "private mode" signaling under NFPA 72 to the fire safety and evacuation plan requirements of Chapter 4. Section 907.5.2.1 has been revised to allow the use of a private mode audible alarm in critical care areas. Section 907.5.2.3 has been revised to allow for the substitution of an audible alarm for a visual alarm in critical care areas.

2015 CODE: 907.2.6 Group I.   A manual fire alarm system that activates the occupant notification system in accordance with Section 907.5 shall be installed in Group I occupancies. An automatic smoke detection system that activates the occupant notification system in accordance with Section 907.5 shall be provided in accordance with Sections 907.2.6.1, 907.2.6.2, and 907.2.6.3.3.

Exceptions:

1. Manual fire alarm boxes in *sleeping units* of Group I-1 and I-2 occupancies shall not be required at *exits* if located at all care providers' control stations or other constantly attended staff locations, provided such stations are visible and continuously accessible and that ~~travel distances~~ the distances of travel required in Section 907.4.2.1 are not exceeded.

2. Occupant notification systems are not required to be activated where private mode signaling installed in accordance with NFPA 72 is approved by the fire code official and staff evacuation responsibilities are included in the fire safety and evacuation plan required by Section 404.

907.2.6, 907.5.2.1, 907.5.2.3 continues

907.2.6, 907.5.2.1, 907.5.2.3

Fire Alarm and Detection Systems for Group I-2 Condition 2 Occupancies

International Code Council®

Visible alarm notification appliance at nurses' control station

907.2.6, 907.5.2.1, 907.5.2.3
continued

907.5.2.1 Audible Alarms. Audible alarm notification appliances shall be provided and emit a distinctive sound that is not to be used for any purpose other than that of a fire alarm.

Exceptions:

1. ~~Visible alarm notification appliances shall be allowed in lieu of audible alarm notification appliances in critical care areas of Group I-2 occupancies.~~ Audible alarm notification appliances are not required in critical care areas of Group I-2 Condition 2 occupancies that are in compliance with Section 907.2.6, Exception 2.

2. A visible alarm notification appliance installed in a nurses' control station or other continuously attended staff location in a Group I-2 Condition 2 suite shall be an acceptable alternative to the installation of audible alarm notification appliances throughout the suite in Group I-2 Condition 2 occupancies that are in compliance with Section 907.2.6, Exception 2.

~~2~~ **3.** Where provided, audible notification appliances located in each occupant evacuation elevator lobby in accordance with Section 3008.10.1 of the *International Building Code* shall be connected to a separate notification zone for manual paging only.

907.5.2.3 Visible Alarms. Visible alarm notification appliances shall be provided in accordance with Sections 907.5.2.3.1 through 907.5.2.3.3.

Exceptions:

1. Visible alarm notification appliances are not required in alterations, except where an existing fire alarm system is upgraded or replaced, or a new fire alarm system is installed.

2. Visible alarm notification appliances shall not be required in exits as defined in Chapter 2.

3. Visible alarm notification appliances shall not be required in elevator cars.

4. Visual alarm notification appliances are not required in critical care areas of Group I-2 Condition 2 occupancies that are in compliance with Section 907.2.6, Exception 2.

CHANGE SIGNIFICANCE: The changes to Sections 907.2.6, 907.5.2.1 and 907.5.2.3 are all intended to permit the use of "private mode" signaling in critical care areas of Group I-2 Condition 2 occupancies when approved by the fire code official and when staff evacuation responsibilities are included in the fire safety and evacuation plan required by Section 404. The use of private mode appliances relies on trained staff to respond to and provide for occupants.

In Section 907.5.2.1 Exception 1, the language changes from the current use of "private mode" as an alternative form of protection to not requiring an audible signaling appliance for critical care areas of I-2 Condition 2 occupancies. However, the substitution of "private mode" still requires an audible alarm notification, which is at much lower decibel levels and is intended to alert staff only. A private mode alarm is not intended to initiate a general evacuation. Exception 2 has been added to allow for a visible alarm notification device located at the nurses' control station or other continuously attended staff location to be an acceptable alternative to the installation of audible alarm notification appliances throughout the suite in Group I-2 Condition 2 occupancies that are in compliance with Section 907.2.6, Exception 2. The rationale is that at a constantly attended location, a visual signaling device will catch the attention of the staff, and they can respond to the alarm in accordance with their training.

The 2012 IFC allowed "private mode" audible signaling as an alternative form of protection in critical care areas while still requiring visual notification signaling. The change to 907.5.2.3 changes the code to not require visual signaling appliances in Group I-2 Condition 2 occupancies. Experience over the past few years has shown that these visual signaling appliances are a distraction in critical care areas such as operating rooms, and they cannot be deactivated in a timely manner to still provide the best care for the patient. Accordingly, they have been eliminated.

NFPA 72, *National Fire Alarm and Signaling Code* Section 3.3.185.1 defines Private Operating Mode as

Audible or visual signaling only to those persons directly concerned with the implementation and direction of emergency action initiation and procedure in the area protected by the fire alarm system. (SIG-NAS)

The code states:

Institutional Group I-2. This occupancy shall include buildings and structures used for medical care on a 24-hour basis for more than five persons who are not capable of self-preservation. This group shall include, but not be limited to, the following:

Foster care facilities
Detoxification facilities
Hospitals
Nursing homes
Psychiatric hospitals.

Occupancy Conditions. Buildings of Group I-2 shall be classified as one of the following occupancy conditions:

Condition 1. This occupancy condition shall include facilities that provide nursing and medical care but do not provide emergency care, surgery, obstetrics, or in-patient stabilization units for psychiatric or detoxification, including, but not limited to nursing homes and foster care facilities.

Condition 2. This occupancy condition shall include facilities that provide nursing and medical care and could provide emergency care, surgery, obstetrics, or inpatient stabilization units for psychiatric or detoxification, including, but not limited to hospitals.

907.2.9.3

Fire Alarm and Detection Systems for Group R College and University Buildings

Fire alarm panel for automatic smoke detection system

CHANGE TYPE: Clarification

CHANGE SUMMARY: The addition of the language "occupancies operated by a college or university for student or staff housing" is intended to clarify this section and the requirement for automatic smoke detection.

2015 CODE: 907.2.9.3 Group R-2 College and University Buildings. An automatic smoke detection system that activates the occupant notification system in accordance with Section 907.5 shall be installed in Group R-2 occupancies operated by a college ~~and~~ or university for student or staff housing ~~buildings~~ in all of the following locations:

1. Common spaces outside of dwelling units and sleeping units.
2. Laundry rooms, mechanical equipment rooms and storage rooms.
3. All interior corridors serving sleeping units or dwelling units.

Exception: An automatic smoke detection system is not required in buildings that do not have interior corridors serving sleeping units or dwelling units and where each sleeping unit or dwelling unit either has a means of egress door opening directly to an exterior exit access that leads directly to an exit or a means of egress door opening directly to an exit.

Required smoke alarms in dwelling units and sleeping units in Group R-2 occupancies operated by a college ~~and~~ or university for student or staff housing shall be interconnected with the fire alarm system in accordance with NFPA 72.

Exceptions: ~~An automatic smoke detection system is not required in buildings that do not have interior corridors serving sleeping units or dwelling units and where each sleeping unit or dwelling unit either has a means of egress door opening directly to an exterior exit access that leads directly to an exit or a means of egress door opening directly to an exit.~~

CHANGE SIGNIFICANCE: This change clarifies the intent of the code so that the section clearly includes those buildings that are operated by a college or university for student or staff housing (regardless of whether the college or university actually owns the building). The difficulty with the 2012 version was in determining how the section applied to off-campus housing that is open to the general public. It provided no guidance in determining a threshold at which a "typical" apartment building becomes subject to the provisions of this section.

Given the number of fires and fatalities in campus housing over the past few years, the code intends that "dormitory style" student housing that is operated by a college or university must comply with this code requirement. In addition, the existing exception has been relocated in the section so that it is properly placed with respect to the paragraph that it applies to. No change has been made to the exception text.

International Code Council®

Campus firewatch (www.campus-firewatch.com) listed the following statistics on their information sheet dated November 16, 2013:

CAMPUS-RELATED FIRE FATALITIES FROM JANUARY 2000 TO NOVEMBER 16, 2013

Occupancy	Deaths	% of total
Off-campus	141	87
Residence Hall	10	6
Greek Housing	10	6
Other	2	1
Total	**163**	**100**

907.2.11.3, 907.2.11.4

Smoke Alarms near Cooking Appliances and Bathrooms

CHANGE TYPE: Addition

CHANGE SUMMARY: This new section provides designers, plan examiners and field inspectors with criteria for locating smoke alarms in relation to cooking appliances and bathrooms. By properly locating smoke alarms, the number of nuisance alarms may be reduced.

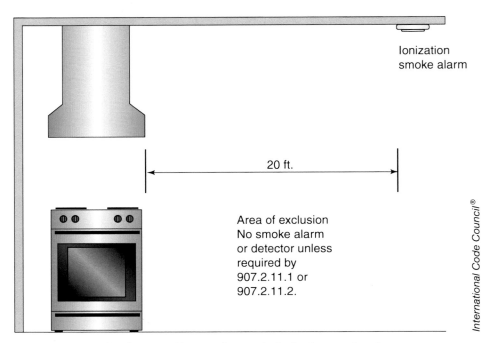

Ionization smoke alarm

20 ft.

Area of exclusion
No smoke alarm
or detector unless
required by
907.2.11.1 or
907.2.11.2.

International Code Council®

Minimum separation from cooking appliances to ionization smoke alarm

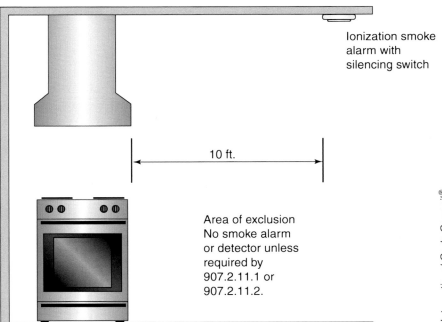

Ionization smoke
alarm with
silencing switch

10 ft.

Area of exclusion
No smoke alarm
or detector unless
required by
907.2.11.1 or
907.2.11.2.

International Code Council®

Minimum separation from cooking appliances to ionization smoke alarm with alarm silencing switch

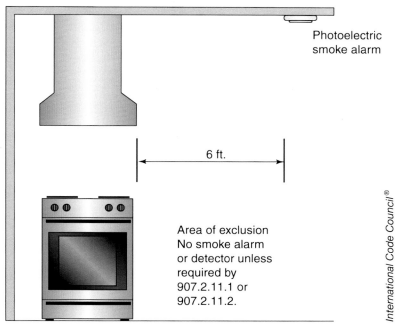

Minimum separation from cooking appliances to photoelectric smoke alarm

2015 CODE: **907.2.11.3 Installation Near Cooking Appliances.** Smoke alarms shall not be installed in the following locations unless this would prevent placement of a smoke alarm in a location required by Sections 907.2.11.1 or 907.2.11.2.

1. Ionization smoke alarms shall not be installed less than 20 feet (6096 mm) horizontally from a permanently installed cooking appliance.

2. Ionization smoke alarms with an alarm-silencing switch shall not be installed less than 10 feet (3048 mm) horizontally from a permanently installed cooking appliance.

3. Photoelectric smoke alarms shall not be installed less than 6 feet (1829 mm) horizontally from a permanently installed cooking appliance.

907.2.11.4 Installation Near Bathrooms. Smoke alarms shall be installed not less than 3 feet (914 mm) horizontally from the door or opening of a bathroom that contains a bathtub or shower unless this would prevent placement of a smoke alarm required by Sections 907.2.11.1 or 907.2.11.2.

907.2.11.3, 907.2.11.4 continues

907.2.11.3, 907.2.11.4 continued

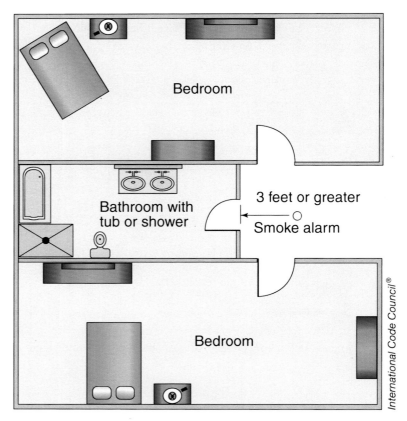

Minimum bathroom separation from smoke alarm

CHANGE SIGNIFICANCE: This change is intended to reduce nuisance alarms attributed to locating smoke alarms in close proximity to cooking appliances and bathrooms where water vapor is produced. The changes are based on the findings in the 2008 NFPA Task Group Report, "Minimum Performance Requirements for Smoke Alarm Detection Technology," February 22, 2008, and are consistent with similar requirements included in Section 29.8.3.4 of the 2010 and 2013 editions of NFPA 72, *National Fire Alarm and Signaling Code.*

Nothing in these sections dictate when a smoke alarm is required; only installation parameters are covered. Specific requirements are in Sections 907.2.11.1 and 907.2.11.2.

CHANGE TYPE: Addition

CHANGE SUMMARY: This new section provides an option for using a smoke detection system in lieu of single-station and multiple-station alarms in Groups R-2, R-3, R-4 and I-1.

2015 CODE: **907.2.11.7 Smoke Detection System.**   Smoke detectors listed in accordance with UL 268 and provided as part of the building's fire alarm system shall be an acceptable alternative to single- and multiple-station smoke alarms and shall comply with the following:

1. The fire alarm system shall comply with all applicable requirements in Section 907.

2. Activation of a smoke detector in a dwelling unit or sleeping unit shall initiate alarm notification in the dwelling unit or sleeping unit in accordance with Section 907.5.2.

3. Activation of a smoke detector in a dwelling unit or sleeping unit shall not activate alarm notification appliances outside of the dwelling unit or sleeping unit, provided that a supervisory signal is generated and monitored in accordance with Section 907.6.6.

CHANGE SIGNIFICANCE: This change adds a section to allow the option of using a smoke detection system as an alternative to single-station and multiple-station alarms. Specifically, it allows smoke detectors listed in accordance with UL 268, Smoke Detectors for Fire Alarm Systems using ionization-type, photoelectric-type or combination detectors using both photoelectric and ionization technologies. UL is now recommending the use of combination detectors because research that has shown that each technology has unique performance characteristics. Ionization responds quicker to a slow smoldering fire, whereas a photoelectric detector responds quicker to a fast flaming fire.

In larger buildings of Group R-2 and 1-1 occupancies, this can be a significant cost savings because the installation and subsequent testing of duplicate devices inside a dwelling unit or sleeping unit are avoided. Addressable smoke detection systems can be easily programmed to alert only the occupants of the dwelling unit or sleeping unit while at the same time maintaining supervision of the circuit integrity. The testing and maintenance of today's addressable systems is more easily accomplished using the sensitivity reporting features of these systems. For jurisdictions that have been allowing this methodology, this change will eliminate the need for approving these systems under the alternative materials and methods section of the code.

907.2.11.7
Smoke Detection System

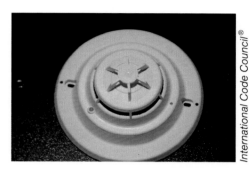

Smoke detector in R-2 occupancy

International Code Council®

907.2.14

Fire Alarm and Detection Systems for Atriums

Atrium connecting more than two stories

CHANGE TYPE: Clarification

CHANGE SUMMARY: This change clarifies that smoke detection in atriums is to be based on the rational analysis prescribed in Section 909.4 and that a generic requirement for installation of smoke detection is not necessarily warranted.

2015 CODE: 907.2.14 Atriums Connecting More Than Two Stories. A fire alarm system shall be installed in occupancies with an atrium that connects more than two stories, with smoke detection ~~installed throughout the atrium~~ <u>in locations required by a rational analysis in Section 909.4 and in accordance with the system operation requirements in Section 909.17</u>. The system shall be activated in accordance with Section 907.5. Such occupancies in Group A, E or M shall be provided with an emergency voice/alarm communication system complying with the requirements of Section 907.5.2.2.

CHANGE SIGNIFICANCE: In the 2003 and 2006 editions, the IFC required smoke detection in atriums as determined by a rational analysis in accordance with Section 909. However, in the 2009 and 2012 editions, a code change was approved that created a new requirement for a smoke detection system in atriums regardless of the rational analysis (Section 907.2.13 [2009] and Section 907.2.14 [2012]).

The change in the 2015 code takes the section back to the original intent, which was that the engineering analysis and the timing of the smoke control system in accordance with Sections 909.4 and 909.17 (smoke control system response time,) respectively, would determine if smoke detectors would be required to maintain a tenable environment for the evacuation or relocation for the occupants of the building. The change maintains the appropriate terminology but returns the requirements to their original intent that the requirements for smoke detection in atriums are unique to each atrium and should be determined by the required rational analysis. In some cases, unique spacing and zoning is required. Also the way in which the system must be initiated may vary based on the unique conditions. The type of detection technology will likely vary. Section 909.12.3 discusses the different modes of automatic activation of the smoke control system.

International Code Council®

CHANGE TYPE: Addition

CHANGE SUMMARY: This new section provides specific criteria regarding smoke detector locations in airport traffic control towers. A different criterion is used depending on whether or not the airport traffic control tower has single or multiple exits and if it is sprinklered.

2015 CODE: 907.2.22 Airport Traffic Control Towers. An automatic smoke detection system that activates the occupant notification system in accordance with Section 907.5 shall be provided in airport traffic control towers in ~~all occupiable and equipment spaces~~ accordance with Sections 907.2.22.1 and 907.2.22.2.

> **Exception:** Audible appliances shall not be installed within the control tower cab.

907.2.22.1 Airport Traffic Control Towers with Multiple Exits and Automatic Sprinklers. Airport traffic control towers with multiple exits and equipped throughout with an automatic sprinkler system in accordance with Section 903.3.1.1, shall be provided with smoke detectors in the following locations.

1. Airport traffic control cab.
2. Electrical and mechanical equipment rooms.
3. Airport terminal radar and electronics rooms.
4. Outside each opening into interior exit stairways.
5. Along the single means of egress permitted from observation levels.
6. Outside each opening into the single means of egress permitted from observation levels.

907.2.22.2 Other Airport Traffic Control Towers. Airport traffic control towers with a single exit or where sprinklers are not installed throughout shall be provided with smoke detectors in the following locations.

1. Airport traffic control cab.
2. Electrical and mechanical equipment rooms.
3. Airport terminal radar and electronics rooms.
4. Office spaces incidental to the tower operation.
5. Lounges for employees, including sanitary facilities.
6. Means of egress.
7. Accessible utility shafts.

CHANGE SIGNIFICANCE: This change correlates the IFC with the NFPA 101, *Life Safety Code*® for air traffic control towers. The change is significant in that it provides a scaled approach to smoke detection based on whether a tower has single or multiple exits and if automatic sprinkler

907.2.22.1, 907.2.22.2 continues

907.2.22.1, 907.2.22.2

Smoke Detection for Airport Traffic Control Towers

Airport traffic control tower with multiple exits

International Code Council®

907.2.22.1, 907.2.22.2 continued

protection is provided. In conjunction with this change to the IFC, the IBC has been changed in Section 412.3 through the addition of specific requirements regarding the arrangement of exits and the requirement of automatic sprinkler systems where an occupied floor is located more than 35 feet above the lowest level of fire department vehicle access (IBC Section 412.3.6).

When airport traffic control towers are not provided with at least two exits or sprinkler protection is not provided throughout, the following additional locations need to be provided with smoke detection:

- Office spaces incidental to the tower operation.
- Lounges for employees, including sanitary facilities.
- Means of egress.
- Accessible utility shafts.

909.4.7
Smoke Control System Interaction

CHANGE TYPE: Addition

CHANGE SUMMARY: This new section requires the analysis of multiple mechanical smoke control systems. Buildings using smoke control systems may have more than one type of smoke control system, and the interactions of these systems must be evaluated in the design.

2015 CODE: **909.4.7 Smoke Control System Interaction.**   The design shall consider the interaction effects of the operation of multiple smoke control systems for all design scenarios.

CHANGE SIGNIFICANCE: The designer must conduct a specific analysis of the interactions of multiple mechanical smoke control systems so that the systems maintain a tenable environment for the evacuation or relocation of occupants.

It is not uncommon for a high-rise building to contain multiple systems, such as an atrium with smoke control, pressurized stairways and hoistway pressurization used as an option for compliance with enclosed elevator lobby requirements. If these three smoke control systems are running simultaneously, they may not work as intended unless the interaction effects are evaluated during the design for all possible design fire scenarios.

A rational analysis performed during the design phase of a project should save time, money, and aggravation during the commissioning and certificate of occupancy phase. If the design does not take into account the interactions of systems, it is during the commissioning phase that the inspector usually discovers that these systems are not working as intended due to conflicting interactions with other smoke control systems.

High-rise building with multiple smoke control systems

909.6.3

Smoke Control Systems—Pressurized Stairways and Elevator Hoistways

CHANGE TYPE: Addition

CHANGE SUMMARY: This section has been added for clarification of the responsibility and authority between the fire code official and the building official in relation to smoke control systems.

2015 CODE: **909.6.3 Pressurized Stairways and Elevator Hoistways.** Where stairways or elevator hoistways are pressurized, such pressurization systems shall comply with Section 909 as smoke control systems, in addition to the requirements of Section 909.21 of this code and Section 909.20 of the *International Building Code.*

CHANGE SIGNIFICANCE: This section in conjunction with Section 909.21 brings into the IFC the requirements for pressurization that previously were only in the IBC. The lack of duplicative requirements in both the IFC and IBC has led to uncertainty on the part of designers as to the appropriate authority for these sections. Further compounding the problem was the need for close coordination of the smoke control design and the fire alarm design. When jurisdictions are not coordinated internally it is sometimes difficult for them to complete the commissioning tests and receive final approvals from all authorities having jurisdiction.

Pressurized stairway and elevator hoistways

CHANGE TYPE: Modification

CHANGE SUMMARY: This modification allows the fire code official the discretion to bypass individual components from the weekly preprogrammed smoke control verification testing. It further requires testing of all bypassed components on a semiannual basis.

2015 CODE: 909.12.1 Verification.   Control systems for mechanical smoke control systems shall include provisions for verification. Verification shall include positive confirmation of actuation, testing, manual override~~,~~ and the presence of power downstream of all disconnects~~, and through a.~~ A preprogrammed weekly test sequence~~,~~ shall report abnormal conditions audibly, visually and by printed report. The preprogrammed weekly test shall operate all devices, equipment, and components used for smoke control.

> **Exception:**   Where verification of individual components tested through the preprogrammed weekly testing sequence will interfere with, and produce unwanted effects to normal building operation, such individual components are permitted to be bypassed from the preprogrammed weekly testing, where approved by the fire code official and in accordance with both the following:
>
> 1. Where the operation of components is bypassed from the preprogrammed weekly test, presence of power downstream of all disconnects shall be verified weekly by a listed control unit.
> 2. Testing of all components bypassed from the preprogrammed weekly test shall be in accordance with Section 909.20.6.

909.12.1, 909.20.6 continues

909.12.1, 909.20.6

Verification of Mechanical Smoke Control Systems

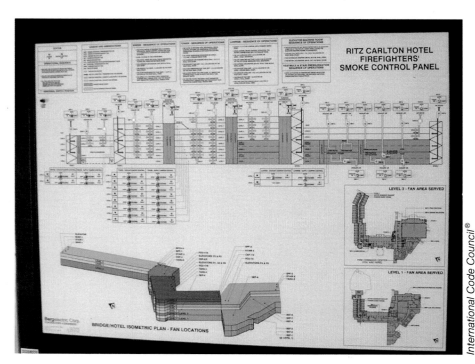

Building with smoke control and verification

International Code Council®

909.12.1, 909.20.6 continued

909.20.6 Components Bypassing Weekly Test. Where components of the smoke control system are bypassed by the preprogrammed weekly test required by Section 909.12.1, such components shall be tested semi-annually. The system shall also be tested under standby power conditions.

CHANGE SIGNIFICANCE: The current provisions require a weekly test of smoke control systems. For many systems, the weekly test requires the introduction of untreated air into the smoke zone. This can be impractical in areas with cold or hot climates and for buildings that require close control of temperature and humidity, such as art museums and similar facilities. The introduction of the untreated air can also result in energy wasted to reheat, re-cool, humidify, or dehumidify the smoke control zone.

The intent of the current code provisions is to provide a means to verify that the required systems will be available when needed. The code requires control units to comply with UL 864, Standard for Control Units and Accessories for Fire Alarm Systems. Thus, all components of the control system will be supervised.

The new exceptions allow components to be bypassed providing a listed control unit monitors the presence of power downstream of all disconnects. This will provide reasonable assurance that power will be available for all smoke control components such as fans, dampers, doors, and windows. In addition, those components that are bypassed must be tested on a semiannual basis.

CHANGE TYPE: Addition

CHANGE SUMMARY: This change provides the option of pressurizing the elevator hoistway in lieu of enclosing the elevator lobby. The entire section has been added to the IFC to facilitate coordination between code officials. Additionally, four exceptions have been added to the pressurization requirements that in effect provide an alternative way for the smoke control system to be designed.

2015 CODE: 909.21 Elevator Hoistway Pressurization Alternative. Where elevator hoistway pressurization is provided in lieu of required enclosed elevator lobbies, the pressurization system shall comply with Sections 909.21.1 through 909.21.11.

909.21.1 Pressurization Requirements. Elevator hoistways shall be pressurized to maintain a minimum positive pressure of 0.10 inch of water (25 Pa) and a maximum positive pressure of 0.25 inch of water (67 Pa) with respect to adjacent occupied space on all floors. This pressure shall be measured at the midpoint of each hoistway door, with all elevator cars at the floor of recall and all hoistway doors on the floor of recall open and all other hoistway doors closed. The pressure differential shall be measured between the hoistway and the adjacent elevator landing. The opening and closing of hoistway doors at each level must be demonstrated during this test. The supply air intake shall be from an outside, uncontaminated source located a minimum distance of 20 feet (6096 mm) from any air exhaust system or outlet.

Exceptions:
1. On floors containing only Group R occupancies, the pressure differential is permitted to be measured between the hoistway and a *dwelling unit* or *sleeping unit*.

909.21 continues

909.21
Elevator Hoistway Pressurization Alternative

Elevator hoistways

909.21 continued

2. Where an elevator opens into a lobby enclosed in accordance with Section 3007.6 or 3008.6 of the *International Building Code*, the pressure differential is permitted to be measured between the hoistway and the space immediately outside the door(s) from the floor to the enclosed lobby.

3. The pressure differential is permitted to be measured relative to the outdoor atmosphere on floors other than the following:

   3.1. The fire floor.

   3.2. The two floors immediately below the fire floor.

   3.3. The floor immediately above the fire floor.

4. The minimum positive pressure of 0.10 inch of water (25 Pa) and a maximum positive pressure of 0.25 inch of water (67 Pa) with respect to occupied floors is not required at the floor of recall with the doors open.

909.21.1.1 Use of Ventilation Systems. Ventilation systems, other than hoistway supply air systems, are permitted to be used to exhaust air from adjacent spaces on the fire floor, two floors immediately below and one floor immediately above the fire floor to the building's exterior where necessary to maintain positive pressure relationships as required in Section 909.21.1 during the operation of the elevator shaft pressurization system.

CHANGE SIGNIFICANCE: For consistency, Section 909.21 has been added into the IFC from the IBC. Additional changes introduce new locations for measuring pressure differentials between pressurized hoistways and surrounding spaces.

The new text in Section 909.21.1 clarifies between which two points the pressure differential gets measured. In general, the intent of the code is to keep smoke out of the hoistway, so the pressure should be measured between the elevator hoistway and the elevator landing/lobby.

Exception 1 allows the pressure to be measured between the hoistway and sleeping or dwelling units in residential buildings, since they are highly compartmented. In high-rise hotels, fires occur most often in the dwelling or sleeping units. Positive pressure in the corridor/hallway outside the units (via leakage through the elevator hoistway doors) will help reduce the smoke migrating from the affected unit.

Exception 2 allows the pressure to be measured between the hoistway and the space on the outside of the smoke barrier that forms the lobby enclosure.

Exception 3 represents the most significant change from previous codes. The code now requires the 0.10-inch water column pressure differential between the hoistway and the floor be met only on the four most critical floors—the floor of fire origin, the two floors immediately below, and one floor immediately above. For all other stories, the pressure differential is allowed to be measured between the hoistway and the outside of the building. The purpose of this requirement is to maintain a slightly positive pressure in the building relative to atmospheric in order to lower the neutral pressure plane in the building, which then reduces the driving force of stack effect. This exception is intended to be permitted to be used in conjunction with Exceptions 1 and 2.

Exception 4 does not require a pressure differential across the open hoistway doors on the level of recall, because the recall floor should not be the floor of fire origin. Hoistway doors on the floor of fire origin will not open if smoke is present because of the smoke detectors that protect it.

Section 909.21.1.1 essentially allows a zone smoke control approach instead of a pressurized hoistway.

910

Smoke and Heat Removal

Mechanical exhaust system for high-piled combustible storage

International Code Council®

CHANGE TYPE: Modification

CHANGE SUMMARY: This section has been extensively rewritten as a result of the work done by the Code Technology Committee and specifically the Roof Vent Study Group. It provides direction on Group F-1 and S-1 occupancies greater than 50,000 square feet of undivided area and high-piled combustible storage. Criteria for using either smoke and heat vents or mechanical smoke removal are provided.

2015 CODE:

SECTION 910
SMOKE AND HEAT REMOVAL

910.1 General. Where required by this code, ~~or otherwise installed,~~ smoke and heat vents or mechanical smoke ~~exhaust~~ removal systems ~~and draft curtains~~ shall conform to the requirements of this section.

~~**Exceptions:**~~

~~1. Frozen food warehouses used solely for storage of Class I and II commodities where protected by an approved automatic sprinkler system.~~

~~2. Where areas of buildings are equipped with early suppression fast-response (ESFR) sprinklers, automatic smoke and heat vents shall not be required within these areas.~~

910.2 Where Required. Smoke and heat vents or a mechanical smoke removal system shall be installed ~~in the roofs of buildings or portions thereof occupied for the uses set forth in~~ as required by Sections 910.2.1 and 910.2.2.

Exceptions:

1. Frozen food warehouses used solely for storage of Class I and II commodities where protected by an approved automatic sprinkler system.

~~2. In occupied portions of a building where the upper surface of the story is not a roof assembly, mechanical smoke exhaust in accordance with Section 910.4 shall be an acceptable alternative.~~

2. Smoke and heat removal shall not be required in areas of buildings equipped with early suppression fast-response (ESFR) sprinklers.

3. Automatic smoke and heat vents shall not be required in areas of buildings equipped with control mode special application sprinklers with a response time index of 50 $(m \cdot S)^{1/2}$ or less that are listed to control a fire in stored commodities with 12 or fewer sprinklers.

910.2.1 Group F-1 or S-1. Smoke and heat vents installed in accordance with Section 910.3 or a mechanical smoke removal system installed in accordance with Section 910.4 shall be installed in buildings and portions thereof used as a Group F-1 or S-1 occupancy having more than 50,000 square feet (4645 m²) of undivided area. In occupied portions of a building equipped throughout with a sprinkler system in accordance with Section 903.3.1.1, where the upper surface of the story is not a roof assembly, a mechanical smoke removal system in accordance with Section 910.4 shall be installed.

Exception: Group S-1 aircraft repair hangars.

910.2.2 High-Piled Combustible Storage. Smoke and heat removal required by Table 3206.2, for buildings and portions thereof containing high-piled combustible ~~stock or rack~~ storage shall be installed in accordance with Section 910.3 in unsprinklered buildings. In buildings and portions thereof containing high-piled combustible storage equipped throughout with an automatic sprinkler system in accordance with Section 903.3.1.1 ~~in any occupancy group when required by Section 3206.7.~~ a smoke and heat removal system shall be installed in accordance with Section 910.3 or 910.4. In occupied portions of a building equipped throughout with an automatic sprinkler system in accordance with Section 903.3.1.1 where the upper surface of the story is not a roof assembly, a mechanical smoke removal system in accordance with Section 910.4 shall be installed.

910.3 Smoke and Heat Vents ~~Design and Installation~~. The design and installation of smoke and heat vents ~~and draft curtains~~ shall be ~~as specified~~ in accordance with Sections 910.3.1 through 910.3.3 ~~910.3.5.2 and Table 910.3~~.

910.3.1 ~~Design~~ Listing and Labeling. Smoke and heat vents shall be listed and labeled to indicate compliance with UL 793 or FM 4430.

910.3.2 Smoke and Heat Vent Locations. Smoke and heat vents shall be located 20 feet (6096 mm) or more from adjacent lot lines and fire walls and 10 feet (3048 mm) or more from fire barriers. Vents shall be uniformly located within the roof in the areas of the building where the vents are required to be installed by Section 910.2, with consideration given to roof pitch, ~~draft curtain location~~, sprinkler location and structural members.

910.3.3 Smoke and Heat Vents Area. The required aggregate area of smoke and heat vents shall be calculated as follows:

For buildings equipped throughout with an automatic sprinkler system in accordance with Section 903.3.1.1:

$$A_{VR} = V/9000 \qquad \text{(Equation 9-4)}$$

where
A_{VR} = the required aggregate vent area (ft²)
V = volume (ft³) of the area that requires smoke removal

910 continues

910 continued

For unsprinklered buildings:

$$A_{VR} = A_{FA}/50$$
(Equation 9-5)

where

A_{VR} = the required aggregate vent area (ft²)
A_{FA} = the area of the floor of the area that requires smoke removal.

Section 910 has been substantially rewritten. For brevity and clarity only portions of the code text are shown. See the 2015 IFC for the complete code text.

CHANGE SIGNIFICANCE: This rewrite of Section 910 updates it to correspond to the latest technical information developed in the United States over the past 20 years. The baseline that this section uses is that the primary purpose of smoke and heat removal is to assist fire-fighting efforts in the salvage and overhaul phase of fireground operations after control of the fire has been achieved by the automatic sprinkler system.

The use of automatic sprinkler systems and automatic smoke and heat vents were developed independently, and over time their interaction has been a concern. Specifically, the possibility that the smoke and heat vents may operate before the automatic sprinkler system can control the fire is of concern.

The ICC Code Technology Committee was the proponent of this change, and in the development of the change they created the Roof Vent Study Group (RVSG). One of the findings of this study group was that a manually activated mechanical smoke removal system could perform the same function as roof vents. Accordingly, this change increases the emphasis and acceptability of mechanical smoke removal systems as an alternative to smoke and heat vents. Mechanical smoke removal systems as prescribed in this change provide fire-rated, grade-level enclosures for the control of the mechanical smoke removal system. This provides greater control of the system for the fire-incident commander and reduces his or her need to place fire fighters on roofs or in other hazardous situations to operate smoke and heat venting systems.

The following is a summary of the changes to Section 910.

- Either automatic roof vents or a manually-activated mechanical smoke removal system are permitted in Group F-1 and S-1 occupancies greater than 50,000 square feet of undivided area. See Section 910.2.1.
- Only roof vents are permitted in buildings with high-piled combustible storage that are not protected by a sprinkler system (i.e., buildings that contain high-piled storage with an area between 2,500 and 12,000 square feet). The rationale for this provision is two-fold: first, that the automatic function of smoke and heat vents performs as intended in a non-sprinklered building; and second, that a mechanical smoke removal system capable of handling temperatures between 1,000°F and 2,000°F cannot be practically provided at a reasonable cost. See Section 910.2.2.
- Provisions for the design of roof vents in buildings protected by a sprinkler system have been modified to require that the area of roof vents provide equivalent venting to that required for the mechanical smoke removal system (two air changes per hour) based on an

assumption that each square foot of vent area will provide 300 cubic feet per minute (cfm) of ventilation. See Section 910.3.3.

- Provisions for the design of roof vents in buildings not protected by a sprinkler system have been simplified to require that the ratio of the area of the vents to the floor area be a minimum of 1:50. The rationale for this revision is that roof vents provided without sprinkler protection will be rare: buildings that contain high-piled storage with an area between 2,500 and 12,000 square feet. Given that this situation will be rare, a complex analysis to determine the required area of roof vents is unnecessary. The ratio of vent area to floor area of 1:50 is conservative based on the present requirements included in the IBC and IFC. See Section 910.3.3.

- Provisions for the design of a manually activated mechanical smoke removal system have been included. The provisions are specific in requiring manual activation only. These provisions require that the mechanical smoke removal system be sized to provide a minimum exhaust rate of two air changes per hour based on the enclosed volume of the building space to be exhausted, without any deductions for the space occupied by storage or equipment. See Section 910.4.

- Since it is anticipated that the mechanical smoke removal system will be activated only after the arrival of fire fighters at the scene (estimated to be 7 minutes or longer after ignition), ceiling temperatures should be reduced sufficiently to allow fans rated for only ambient temperatures to be used. See Section 910.4.2.

- The existing provisions for the design of a mechanical smoke removal system indicate that the electrical power supply for the system is to be wired ahead of the main building disconnect for increased reliability and to facilitate fire-fighting operations. No changes have been made to this sub-section. See Section 910.4.6.

- The provisions pertaining to draft curtains have been removed because research conducted by Factory Mutual Research Corporation (FMRC) in 1994 and research conducted at UL in 1997/1998 demonstrated that draft curtains affect the sequence of operation of sprinklers and may have an adverse effect on sprinkler operation.

913.2.2

Electric Circuits Supplying Fire Pumps

CHANGE TYPE: Addition

CHANGE SUMMARY: This new provision references UL Standard 2196, which provides for survivability of fire pump power-supply wiring.

2015 CODE: <u>**913.2.2 Circuits Supplying Fire Pumps.** Cables used for survivability of circuits supplying fire pumps shall be listed in accordance with UL 2196. Electrical circuit protective systems shall be installed in accordance with their listing requirements.</u>

CHAPTER 80
REFERENCED STANDARDS

UL
<u>2196-2001 Tests for Fire-Resistive Cables, with revisions through March 2012 . . . . 913.2.2</u>

CHANGE SIGNIFICANCE: In a fire situation, the reliability and integrity of a fire pump is critical. The 2013 edition of NFPA 20, Standard for the Installation of Stationary Pumps for Fire Protection is referenced by the IFC. One of the provisions in the standard requires protection of the power-supply wiring to electric fire pumps to assure integrity of certain critical circuits.

NFPA 70, *National Electric Code*® does not specify the applicable test standard within the mandatory provisions of the code, but it does recognize electrical circuit protective systems as an alternative to listed cables. NFPA 70 states that one of the following methods must be used to protect the electrical system:

1. Encased in a minimum of 2 inches of concrete.
2. Protected by a 2-hour fire-resistance-rated assembly.
3. Listed electrical circuit protective system with a minimum fire-resistance rating of 2 hours.

The new Section 913.2.2 requires that the power-supply cables be listed to UL 2196, Tests for Fire-Resistive Cables. This standard evaluates the fire-resistive performance of electrical cables during and after a fire test exposure and hose stream test.

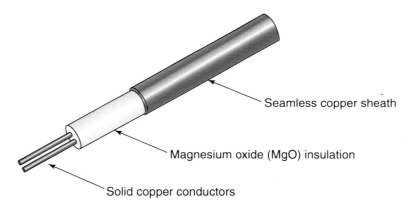

Seamless copper sheath

Magnesium oxide (MgO) insulation

Solid copper conductors

Pyrotenax MI cable—2-hour fire-rated cable *(Graphic courtesy of Pentair Thermal Management, Houston, TX)*

Systems complying with UL 2196 are likely to include a combination of fire-resistive cables, support and mounting details, and a description of the building assembly to which they are to be attached, such as a minimum 2-hour fire-resistance-rated concrete or masonry wall or concrete floor. The fire rating of the wall or floor-ceiling assembly is intended to be equal to or greater than the rating of the electrical circuit integrity system.

Testing under UL 2196 is conducted in accordance with the intended installation method as prescribed by the manufacturer's instructions. Fire-resistive cables designed to be installed without any other fire-resistive barrier are tested as they would be installed. Fire-resistive cables incorporating a fire-resistive jacket are tested without any barrier. If the fire-resistive cables are intended to be installed within a non-fire-resistive barrier, they are tested with the non-fire-resistive barrier included as part of the test specimen. In this case, the enclosure could limit normal ventilation and affect the test results.

Electric cables relying on a fire-resistive barrier or a fire-resistive assembly would be tested as part of that assembly, even though the actual installation in the field is an assembly of components. Therefore, electrical circuit protective systems must be installed according to the listing requirements and manufacturer's instructions in order for the listing for the system to be maintained. Listing requirements can be found within the UL certifications available in the Online Certification Directory at www.ul.com/database.

915

Carbon Monoxide Detection

CHANGE TYPE: Modification

CHANGE SUMMARY: The requirements for carbon monoxide detection have been completely rewritten to clarify the provisions, relocated to a new Section 915, and expanded to address classrooms in Group E occupancies.

2015 CODE: ~~**908.7 Carbon Monoxide Alarms.** Group I or R occupancies located in a building containing a fuel-burning appliance or in a building which has an attached garage shall be equipped with single-station carbon monoxide alarms. The carbon monoxide alarms shall be listed as complying with UL 2034 and be installed and maintained in accordance with NFPA 720 and the manufacturer's instructions. An open parking garage, as defined in Chapter 2 of the *International Building Code*, or an enclosed parking garage ventilated in accordance with Section 404 of the *International Mechanical Code* shall not be considered an attached garage.~~

~~**Exception:** Sleeping units or dwelling units which do not themselves contain a fuel-burning appliance or have an attached garage, but which are located in a building with a fuel-burning appliance or~~

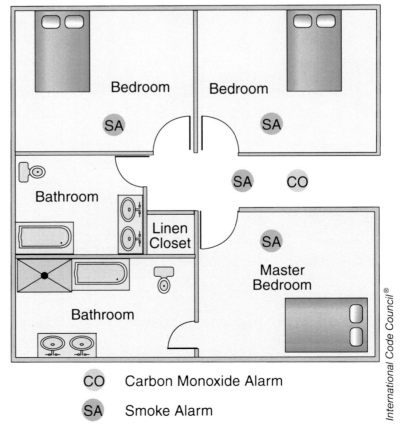

CO   Carbon Monoxide Alarm

SA   Smoke Alarm

The carbon monoxide alarm can be located outside of the sleeping units when there are no fuel-burning appliances within the sleeping units.

International Code Council®

an attached garage, need not be equipped with single-station carbon monoxide alarms provided that:

1. The sleeping unit or dwelling unit is located more than one story above or below any story which contains a fuel-burning appliance or an attached garage;
2. The sleeping unit or dwelling unit is not connected by duct work or ventilation shafts to any room containing a fuel-burning appliance or to an attached garage; and
3. The building is equipped with a common area carbon monoxide alarm system.

908.7.1 Carbon Monoxide Detection Systems. Carbon monoxide detection systems, which include carbon monoxide detectors and audible notification appliances, installed and maintained in accordance with this section for carbon monoxide alarms and NFPA 720 shall be permitted. The carbon monoxide detectors shall be *listed* as complying with UL 2075.

Combination carbon monoxide and smoke alarm

SECTION 915
CARBON MONOXIDE DETECTION

915.1 General. Carbon monoxide detection shall be installed in new buildings in accordance with Sections 915.1.1 through 915.6. Carbon monoxide detection shall be installed in existing buildings in accordance with Section 1103.9.

915.1.1 Where Required. Carbon monoxide detection shall be provided in Group I-1, I-2, I-4 and R occupancies and in classrooms in Group E occupancies in the locations specified in Section 915.2 where any of the conditions in Sections 915.1.2 through 915.1.6 exist.

915.1.2 Fuel-Burning Appliances and Fuel-Burning Fireplaces. Carbon monoxide detection shall be provided in dwelling units, sleeping units and classrooms that contain a fuel-burning appliance or a fuel-burning fireplace.

915.1.3 Forced Air Furnaces. Carbon monoxide detection shall be provided in dwelling units, sleeping units and classrooms served by a fuel-burning, forced air furnace.

Exception: Carbon monoxide detection shall not be required in dwelling units, sleeping units and classrooms where carbon monoxide detection is provided in the first room or area served by each main duct leaving the furnace, and the carbon monoxide alarm signals are automatically transmitted to an approved location.

915.1.4 Fuel-Burning Appliances Outside of Dwelling Units, Sleeping Units and Classrooms. Carbon monoxide detection shall be provided in dwelling units, sleeping units and classrooms located in buildings that contain fuel-burning appliances or fuel-burning fireplaces.

915 continues

915 continued

Exceptions:

1. Carbon monoxide detection shall not be required in dwelling units, sleeping units and classrooms if there are no communicating openings between the fuel-burning appliance or fuel-burning fireplace and the dwelling unit, sleeping unit or classroom.

2. Carbon monoxide detection shall not be required in dwelling units, sleeping units and classrooms if carbon monoxide detection is provided in one of the following locations:

   2.1. In an approved location between the fuel-burning appliance or fuel-burning fireplace and the dwelling unit, sleeping unit or classroom.

   2.2. On the ceiling of the room containing the fuel-burning appliance or fuel-burning fireplace.

915.1.5 Private Garages. Carbon monoxide detection shall be provided in dwelling units, sleeping units and classrooms in buildings with attached private garages.

Exceptions:

1. Carbon monoxide detection shall not be required where there are no communicating openings between the private garage and the dwelling unit, sleeping unit or classroom.

2. Carbon monoxide detection shall not be required in dwelling units, sleeping units and classrooms located more than one story above or below a private garage.

3. Carbon monoxide detection shall not be required where the private garage connects to the building through an open-ended corridor.

4. Where carbon monoxide detection is provided in an approved location between openings to a private garage and dwelling units, sleeping units or classrooms, carbon monoxide detection shall not be required in the dwelling units, sleeping units or classrooms.

915.1.6 Exempt Garages. For determining compliance with Section 915.1.5, an open parking garage complying with Section 406.5 of the *International Building Code* or an enclosed parking garage complying with Section 406.6 of the *International Building Code* shall not be considered a private garage.

915.2 Locations. Where required by Section 915.1.1, carbon monoxide detection shall be installed in the locations specified in Sections 915.2.1 through 915.2.3.

915.2.1 Dwelling Units. Carbon monoxide detection shall be installed in dwelling units outside of each separate sleeping area in the immediate vicinity of the bedrooms. Where a fuel-burning appliance is located within a bedroom or its attached bathroom, carbon monoxide detection shall be installed within the bedroom.

915.2.2 Sleeping Units. Carbon monoxide detection shall be installed in sleeping units.

> **Exception:** Carbon monoxide detection shall be allowed to be installed outside of each separate sleeping area in the immediate vicinity of the sleeping unit where the sleeping unit or its attached bathroom does not contain a fuel-burning appliance and is not served by a forced air furnace.

915.2.3 Group E Occupancies. Carbon monoxide detection shall be installed in classrooms in Group E occupancies. Carbon monoxide alarm signals shall be automatically transmitted to an on-site location that is staffed by school personnel.

> **Exception:** Carbon monoxide alarm signals shall not be required to be automatically transmitted to an on-site location that is staffed by school personnel in Group E occupancies with an occupant load of 30 or less.

915.3 Detection Equipment. Carbon monoxide detection required by Sections 915.1 through 915.2.3 shall be provided by carbon monoxide alarms complying with Section 915.4 or with carbon monoxide detection systems complying with Section 915.5.

915.4 Carbon Monoxide Alarms. Carbon monoxide alarms shall comply with Sections 915.4.1 through 915.4.3.

915.4.1 Power Source. Carbon monoxide alarms shall receive their primary power from the building wiring where such wiring is served from a commercial source, and when primary power is interrupted, shall receive power from a battery. Wiring shall be permanent and without a disconnecting switch other than that required for overcurrent protection.

> **Exception:** Where installed in buildings without commercial power, battery-powered carbon monoxide alarms shall be an acceptable alternative.

915.4.2 Listings. Carbon monoxide alarms shall be listed in accordance with UL 2034.

915.4.3 Combination Alarms. Combination carbon monoxide/smoke alarms shall be an acceptable alternative to carbon monoxide alarms. Combination carbon monoxide/smoke alarms shall be listed in accordance with UL 2034 and UL 217.

915.5 Carbon Monoxide Detection Systems. Carbon monoxide detection systems shall be an acceptable alternative to carbon monoxide alarms and shall comply with Sections 915.5.1 through 915.5.3.

915.5.1 General. Carbon monoxide detection systems shall comply with NFPA 720. Carbon monoxide detectors shall be listed in accordance with UL 2075.

915 continues

915 continued

915.5.2 Locations.   Carbon monoxide detectors shall be installed in the locations specified in Section 915.2. These locations supersede the locations specified in NFPA 720.

915.5.3 Combination Detectors.   Combination carbon monoxide/smoke detectors installed in carbon monoxide detection systems shall be an acceptable alternative to carbon monoxide detectors, provided they are listed in accordance with UL 2075 and UL 268.

915.6 Maintenance.   Carbon monoxide alarms and carbon monoxide detection systems shall be maintained in accordance with NFPA 720. Carbon monoxide alarms and carbon monoxide detectors that become inoperable or begin producing end-of-life signals shall be replaced.

SECTION 202
GENERAL DEFINITIONS

Private Garage.   A building or portion of a building in which motor vehicles used by the tenants of the building or buildings on the premises are stored or kept, without provisions for repairing or servicing such vehicles for profit.

1103.9 Carbon Monoxide Alarms.   Existing Group ~~I~~ I-1, I-2, I-4, and ~~or~~ R occupancies ~~located in a building containing a fuel-burning appliance or a building which has an attached garage~~ shall <u>be provided with</u> ~~be equipped with single-station~~ carbon monoxide alarms <u>in accordance with Section 915, except that the carbon monoxide alarms shall be allowed to be solely battery powered.</u>

CHANGE SIGNIFICANCE: The requirements for carbon monoxide (CO) detection have been relocated and completely rewritten. The relocation is intended to remove CO detection requirements from under the heading of "Emergency Alarms" in Section 908. The new Section 915 is a standalone section dealing entirely with CO detection.

In the 2012 IFC, the installation of CO alarms applied to all Group I and R occupancies where fuel-burning appliances or attached garages are present. In the 2015 IFC, Group I-3 has been removed and classrooms of Group E have been added. CO detection is only required in Groups I-1, I-2, I-4, and R occupancies, and classrooms in a Group E occupancy.

Sections 915.1.2 through 915.1.6 list the particular criteria that require the installation of CO detection in those occupancies. Generally speaking, CO detection is required when the following potential sources of CO exist:

- Fuel-burning appliances in the space or building;
- A fuel-burning fireplace in the space or building;
- A fuel-burning, forced air furnace;
- An attached private garage.

There are several exceptions to the above-listed features, but if none of these potential CO sources exist, then CO detectors are not required. These criteria provide guidance for determining the need for CO detection.

Additionally a definition has been added for the term "private garage." This includes any garage, other than a repair garage, where vehicles are parked by tenants or occupants of the building. This definition would cover a private parking space at an apartment as well as a parking area at a business office. The concern is that CO from vehicles in an attached garage can be introduced into a building, in which case CO detection is required. However, open parking garages complying with IBC Section 406.5 and enclosed parking garage complying with IBC Section 406.6 have construction features in place to prevent unsafe levels of CO from being introduced into the building.

CO detection required by Section 915.1.5 does not apply if the private garage is an

- open parking garage complying with IBC 406.5, or
- enclosed parking garage complying with IBC 406.6.

These private garages will contain a mechanical ventilation system.

Section 915.1.4 covers situations where dwelling units or sleeping units do not contain a fuel-burning appliance, but such an appliance is included in the common area of the building. For example, consider a multi-story hotel that has all-electric HVAC in the sleeping units, but a fireplace in the lobby, fuel-burning forced air heating in the common areas and a fuel-burning boiler in an equipment room. Here, a few strategically located CO detectors will provide a reasonable level of protection for the sleeping units and dwelling units.

Section 915.1.4, Exception 1 covers situations where CO emanating from the fuel-burning appliance has no direct path to a dwelling unit or sleeping unit, such as a water heater in an equipment room that only has access from the exterior of the building. An interior door between this equipment room and a dwelling unit, even if it is self-closing, would not allow this exception to be used.

Section 915.1.4, Exception 2 allows for the installation of one or more CO detectors between the fuel-burning appliances and the nearest dwelling unit or sleeping unit, or on the ceiling of the room in which a fuel burning appliance is located. CO detectors are required only where there are communicating openings, which could include ducts, concealed spaces, interior hallways and stairways between the fuel-burning appliance and the dwelling unit or sleeping unit that would allow air flow from the appliance to the dwelling unit or sleeping unit.

Section 915.1.5 requires CO detectors to be provided when the building has an attached private garage, other than an open parking garage or enclosed parking garage that contain mechanical ventilation systems. Exception 3 allows the elimination of CO detection when the private garage is attached to the building by an open-ended corridor, commonly referred to as a breeze way.

Section 915.2 describes the locations where CO detection is to be provided. In some cases this differs from the requirements in NFPA 720, Standard for the Installation of Carbon Monoxide (CO) Detection and Warning Equipment. The differences are intentional and provide adequate protection for detection of CO. CO detection is required in the following locations:

- Outside of each sleeping area, but in the immediate vicinity of the bedrooms in a dwelling unit.

915 continues

915 continued

- Within each sleeping area of a sleeping unit, or it can be located outside of the sleeping unit but in the immediate vicinity of multiple sleeping units when the sleeping units do not contain fuel-burning appliances.
- Within each classroom of Group E occupancies.

A dwelling unit is a single unit providing complete, independent living facilities for one or more persons, including permanent provisions for living, sleeping, eating, cooking and sanitation (IFC Section 202).

A sleeping unit is a room or space in which people sleep. Sleeping units can also include permanent provisions for living, eating, and either sanitation or kitchen facilities but not both. Such rooms and spaces that are part of a dwelling unit are not considered sleeping units (IFC Section 202).

Once the determination is made that CO detection is required, it is the building owner's option whether CO alarms are installed or a CO detection system is installed. The CO alarms are addressed in Section 915.4, and CO detection systems are addressed in Section 915.5. The code now provides distinction between CO alarms and a CO detection system. This is similar to the concept of smoke alarms and smoke detection systems. Each CO detector is listed to a different UL standard depending on whether it is to be used as a stand-alone or interconnected detector (UL 2034) or a detector that is part of a CO detection system (UL 2075).

Function	CO Alarms	CO Detection System
Listing standard for the CO detector	UL 2034	UL 2075, but detection performance still complies with UL 2034
Power supply	Building power with battery backup, or can be battery power only where commercial power is not available	Building power with battery backup
Requires a control panel for full functionality	No	Yes
Can be combination CO and smoke detector	YES, but listing for smoke detection is UL 217	Yes, but listing for smoke detection is UL 268
Maintained according to NFPA 720	Yes	Yes

Finally, Section 1103.9 has been revised to correlate the requirements for CO detection in both new construction and existing buildings. There is a retroactive requirement for the installation of CO alarms in existing Group I-1, I-2, I-4, and R occupancies. In general, the CO detection requirements for existing occupancies are identical to those for new construction, except the detection devices are allowed to be battery powered. The retroactive requirement does not apply to classrooms in Group E occupancies.

CHANGE TYPE: Modification

CHANGE SUMMARY: The chapter has been reformatted with the provisions for egress requirements from a space or story being consolidated into a new Section 1006 and a new Section 1007.

2015 CODE: The changes to the code text are too extensive to reprint; please refer to the code text in the 2015 IFC and the *2015 IFC Code Changes Resource Collection* available through the International Code Council.

SECTION ~~1015~~ 1006 NUMBERS OF EXITS AND EXIT ACCESS DOORWAYS

~~**SECTION 1021 NUMBER OF EXITS AND EXIT CONFIGURATION**~~

SECTION 1007 EXIT AND EXIT ACCESS DOORWAY CONFIGURATION

SECTION ~~1007~~ 1009 ACCESSIBLE MEANS OF EGRESS

(Renumbering and formatting affect most remaining sections.)

CHANGE SIGNIFICANCE: The egress provisions of Chapter 10 have been reformatted. Predominately this change involves the combining, reformatting, and relocating the requirements from the previous Sections 1015 and 1021 to the general sections at the front of the chapter. Previously, both Section 1015 and Section 1021 contained provisions addressing the number of means of egress and their arrangement. Section 1015 dealt with the required number and arrangement of the means of egress from an individual room or space, while Section 1021 addressed the requirements for the individual story or overall building. Because the relationship of and the distinction between these two sections was not always clear, it led to confusion or unnecessary difficulty in applying the requirements for some code users.

Combining the two sections along with some other provisions into a single location results in making the provisions easier to understand and apply and therefore result in more uniform application of the code. The reformatting allows the requirements to be placed within technical context and clarifies whether the provisions apply to a room or a story. Provisions that were very similar have been combined, so the distinction is clearer or eliminated to reduce confusion. The changes also allow for the new Section 1006 to address the number of means of egress that are required and then have the new Section 1007 address the arrangement or separation of those egress routes.

Relocating these provisions to the front or more general portion of the chapter allows a natural sequential flow of sections based on the occupant load. Section 1004 begins by providing the means to determine the occupant load. Section 1005 then establishes the width requirements for the egress path based on that occupant load. At that point, the two new sections follow, with Section 1006 establishing the number of means of egress required and Section 1007 addressing how those multiple egress paths must be arranged.

Chapter 10
Means of Egress

Interior exit stairway

Egress door to interior exit stairway

1004.1.1

Cumulative Occupant Loads

CHANGE TYPE: Modification

CHANGE SUMMARY: The determination of the cumulative design occupant load for intervening spaces, adjacent levels and adjacent stories has been clarified and combined into a single section. A subsection has been added to address egress from adjacent stories to clarify that the number of occupants from adjacent stories are not added together unless there is a convergence of egress at an intermediate level by occupants leaving a story from above and below that point.

2015 CODE: 1004.1.1 Cumulative Occupant Loads. Where the path of egress travel includes intervening rooms, areas or spaces, cumulative occupant loads shall be determined in accordance with this section.

1004.1.1.1 Intervening Spaces <u>or Accessory Areas</u>. Where occupants egress from one <u>or more</u> room<u>s</u>, area<u>s</u> or space<u>s</u> through ~~another~~ <u>others</u>, the design occupant load shall be <u>the combined occupant load of interconnected accessory or intervening spaces. Design of egress path capacity shall</u> be based on the cumulative <u>portion of</u> occupant loads of all rooms, areas or spaces to that point along the path of egress travel.

1004.1.1.2 Adjacent Levels <u>for Mezzanines</u>. <u>That portion of </u>the *occupant load* of a mezzanine ~~or story~~ with <u>required </u>egress through a room, area or space on an adjacent level shall be added to the occupant load of that room, area or space.

<u>1004.1.1.3 Adjacent Stories.</u> <u>Other than for the egress components designed for convergence in accordance with Section 1005.6, the occupant load from separate stories shall not be added.</u>

CHANGE SIGNIFICANCE: A major portion of the revisions here have been made to help clarify how the occupant load of a space that passes through another space is viewed when determining both the number of exits and also the capacity (or width) of the egress system. Some of this confusion has crept into the code over the past few editions as changes were made to address items such as exit access stairways as opposed to exit

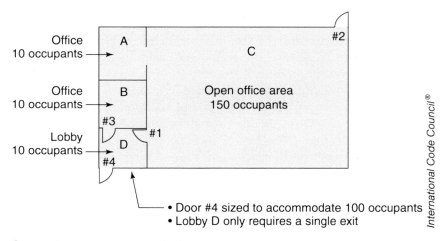

Cumulative occupant loads for intervening spaces

stariways. The intent of these provisions is to emphasize that rooms that share an egress path must be looked at from the aggregate occupant load in order to establish many of the minimum egress requirements and that each path of egress travel must be designed so the capacity of that path is capable of serving the accumulated occupant load that egress along that portion of the path.

Perhaps the biggest technical change is found within Section 1004.1.1.1, and yet to many users it will simply be a clarification. That this change can be viewed so differently depending upon the situation is the reason for this overall change. The first sentence of Section 1004.1.1.1 indicates that where occupants egress from one space through another that the "design occupant load" is the combined or aggregate of the various interconnected or inter-vening spaces. This overall occupant load will be used to establish many of the minimum requirements such as the number of exits or means of egress that must be provided from the overall space, whether doors must swing in the direction of egress, and items such as the minimum component width of 36 inches or 44 inches for stairs or corridors. The second sentence in the code text indicates that it is only the egress width/capacity that is based on the accumulated occupants along that path of travel and that it is not for items such as the number of means of egress.

The purpose of all of these changes is to reinforce the concept that the occupant load is assigned to each occupied area individually. Where there are intervening rooms, each area must be considered both individually and in the aggregate with the other interconnected occupied portions of the exit access in order for the code official to determine the number of means of egress and width of the exit access. Portions of the occupant load are accumulated along the egress path to determine the capacity of individual egress elements along those paths. But once occupants from one area make a choice and head out along one of several independent paths of egress travel, their occupant load is not added to some other area when the code official is determining how many paths of travel would be required from that different area.

Section 1004.1.1.2 recognizes that mezzanines may have independent egress similar to what is typical for a story. If the mezzanine occupants do not egress through the room or area it is a part of, then the occupant load is not added to the main room. If all of the occupants of a mezzanine must egress down through the main room, then their occupant load must be added to the main room or area. If the people on the mezzanine have an option of egress paths such as one independent exit and the other through the room below, then the occupant load is split between the available paths, and the code official must add the portion of the occupants that egress through the room below to the occupant load of that space.

Table 1004.1.2

Occupant Load Factors

CHANGE TYPE: Modification

CHANGE SUMMARY: This change has revised the mercantile occupant load factor and created one factor for all floors.

2015 CODE:

TABLE 1004.1.2 Maximum Floor Area Allowances Per Occupant

Function Of Space	Occupant Load Factor[a]
Mercantile	60 gross
~~Areas on other floors~~	~~60 gross~~
~~Basement and grade floor areas~~	~~30 gross~~
Storage, stock, shipping areas	300 gross

For SI: 1 square foot = 0.0929 m².

a. Floor area in square feet per occupant.

(Remaining portions of table not shown are unchanged.)

CHANGE SIGNIFICANCE: For years the code has provided two occupant load factors for retail spaces, with a smaller occupant load factor being used for grade floors and basements and a larger occupant load factor for other levels. Unfortunately, the code was never clear if the grade floor requirement was applicable to the first floor of the retail space or if it was applicable only to floors that were located at grade level.

The change will allow all retail spaces regardless of which floor level or what type of merchandise to use a single factor of 60 square feet per occupant. When the previous factors were placed into the code they were based on multi-story single operator buildings such as large department stores. Many of these had lower density uses such as furniture or housewares on the upper levels while the spaces on the entry levels were used for higher density sales areas or for things like a bargain basement. Most retail facilities now are not constructed in taller multi-story facilities but instead tend to use larger floor areas and fewer stories. With the changes

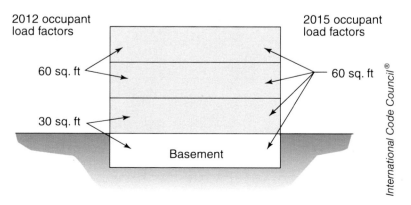

Group M occupancy

2012 occupant load factors — 60 sq. ft — 30 sq. ft

2015 occupant load factors — 60 sq. ft

Basement

International Code Council®

Occupant load factor—mercantile

in retail display and merchandising, the revised code will make it easier for the requirements to be applied, since all floor levels will use the same occupant load factor. The use of the 60 square foot per occupant (based on the gross area) factor matches what was previously accepted for retail sale areas "on other floors." This factor was felt to be a more reasonable number given today's retail environment and that much of the floor area is covered with display cases and counters.

1006, 1007

Numbers of Exits and Exit Access Doorways

CHANGE TYPE: Modification

CHANGE SUMMARY: This modification has consolidated the egress requirements for rooms and spaces along with those for stories into a single location. It has also created a single section to deal with the number of exits (Section 1006) and a separate section (Section 1007) to deal with the arrangement and separation requirements.

2015 CODE: Because this code change affected substantial portions of Chapter 10, the entire code change text is too extensive to be included here. Refer to Code Changes E-1, E-7, E-111, E-127, E-132, and E-136 in the *2015 IBC Code Changes Resource Collection* for the complete text and history of the code change.

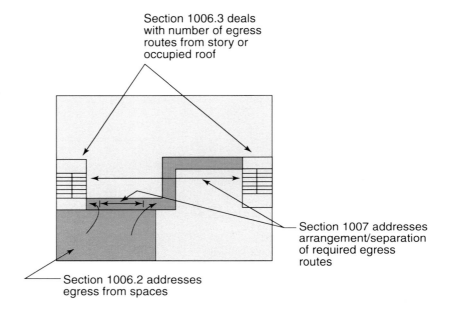

Section 1006.3 deals with number of egress routes from story or occupied roof

Section 1007 addresses arrangement/separation of required egress routes

Section 1006.2 addresses egress from spaces

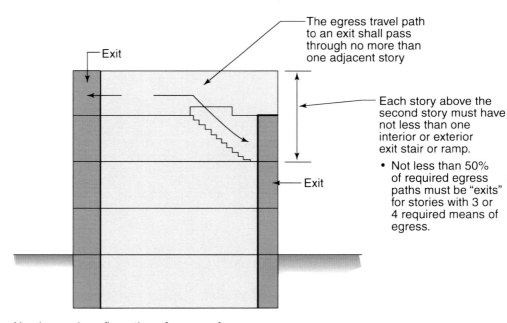

The egress travel path to an exit shall pass through no more than one adjacent story

Exit

Exit

Each story above the second story must have not less than one interior or exterior exit stair or ramp.

• Not less than 50% of required egress paths must be "exits" for stories with 3 or 4 required means of egress.

International Code Council®

Number and configuration of means of egress

~~TABLE 1014.3~~ ~~Common Path of Egress Travel~~

~~OCCUPANCY~~	~~WITHOUT SPRINKLER SYSTEM (feet)~~		~~WITH SPRINKLER SYSTEM (feet)~~
	~~Occupant Load~~		
	~~OL ≤ 30~~	~~OL > 30~~	
~~B, S^d~~	~~100~~	~~75~~	~~100^a~~
~~U~~	~~100~~	~~75~~	~~75^a~~
~~F~~	~~75~~	~~75~~	~~100^a~~
~~H-1, H-2, H-3~~	~~Not Permitted~~	~~Not Permitted~~	~~25^a, g~~
~~R-2~~	~~75~~	~~75~~	~~125^b~~
~~R-3^e~~	~~75~~	~~75~~	~~125^b~~
~~I-3~~	~~100~~	~~100~~	~~100^a~~
~~All others^c, f~~	~~75~~	~~75~~	~~75^a~~

~~For SI: 1 foot = 304.8 mm.~~

~~a. Buildings equipped throughout with an automatic sprinkler system in accordance with Section 903.3.1.1.~~

~~b. Buildings equipped throughout with an automatic sprinkler system in accordance with Section 903.3.1.1 or 903.3.1.2. See Section 903 for occupancies where automatic sprinkler systems are permitted in accordance with Section 903.3.1.2.~~

~~c. For a room or space used for assembly purposes having fixed seating, see Section 1028.8.~~

~~d. The length of a common path of egress travel in a Group S-2 open parking garage shall not be more than 100 feet (30 480 mm).~~

~~e. The length of a common path of egress travel in a Group R-3 occupancy located in a mixed occupancy building.~~

~~f. For the distance limitations in Group I-2, see Section 407.4.~~

~~Occupancy~~	~~Maximum Occupant Load~~
~~A, B, E, F, M, U~~	~~49~~
~~H-1, H-2, H-3~~	~~3~~
~~H-4, H-5, I-1, I-2, I-3, I-4, R~~	~~10~~
~~S~~	~~29~~

~~**1014.3 Common Path of Egress Travel.**~~ ~~The common path of egress travel shall not exceed the common path of egress travel distances in Table 1014.3.~~

SECTION 1006
NUMBERS OF EXITS AND EXIT ACCESS DOORWAYS

1006.1 General. The number of exits or exit access doorways required within the means of egress system shall comply with the provisions of Section 1006.2 for spaces, including mezzanines, and Section 1006.3 for stories.

1006.2 Egress from Spaces. Rooms, areas or spaces, including mezzanines, within a story or basement shall be provided with the number of exits or access to exits in accordance with this section.

1006, 1007 continues

1006, 1007 continued

1006.2.1 Egress Based on Occupant Load and Common Path of Egress Travel Distance. Two exits or exit access doorways from any space shall be provided where the design occupant load or the common path of egress travel distance exceeds the values listed in Table 1006.2.1.

Exceptions:

1. In Group R-2 and R-3 occupancies, one *means of egress* is permitted within and from individual dwelling units with a maximum *occupant load* of 20 where the dwelling unit is equipped throughout with an *automatic sprinkler system* in accordance with Section 903.3.1.1 or 903.3.1.2 and the common path of egress travel does not exceed 125 feet (38 100 mm).

2. Care suites in Group I-2 occupancies complying with Section 407.4.

1006.3 ~~1021.3.1~~ ~~Access to Exits at Adjacent Levels.~~ Egress from Stories or Occupied Roofs. The means of egress system serving any story or occupied roof shall be provided with the number of exits or access to exits based on the aggregate occupant load served in accordance with this section. ~~Access to exits at other levels shall be by stairways~~

TABLE 1006.2.1   **Spaces with One Exit or Exit Access Doorway**

OCCUPANCY	MAXIMUM OCCUPANT LOAD OF SPACE	MAXIMUM COMMON PATH OF EGRESS TRAVEL DISTANCE (feet)		
		WITHOUT SPRINKLER SYSTEM (feet)		WITH SPRINKLER SYSTEM (feet)
		Occupant Load		
		OL ≤ 30	OL > 30	
A[c], E, M	49	75	75	75[a]
B	49	100	75	100[a]
F	49	75	75	100[a]
H-1, H-2, H-3	3	NP	NP	25[b]
H-4, H-5	10	NP	NP	75[b]
I-1, I-2[d], I-4	10	NP	NP	75[a]
I-3	10	NP~~100~~	NP~~100~~	100[a]
R-1	10	NP~~75~~	NP~~75~~	75[a]
R-2	10	NP~~75~~	NP~~75~~	125[a]
R-3[e]	10	NP~~75~~	NP~~75~~	125[a]
R-4[e]	10	75	75	125[a]
S[f]	29	100	75	100[a]
U	49	100	75	75[a]

For SI: 1 foot = 304.8 mm.

NP = Not Permitted.

a. Buildings equipped throughout with an automatic sprinkler system in accordance with Section 903.3.1.1, or 903.3.1.2. See Section 903 for occupancies where automatic sprinkler systems are permitted in accordance with Section 903.3.1.2.

b. Group H occupancies equipped throughout with an *automatic sprinkler system* in accordance with Section 903.2.5.

c. For a room or space used for assembly purposes having fixed seating, see Section 1029.8.

d. For the travel distance limitations in Group I-2, see Section 407.4.

e. The length of common path of egress travel distance in a Group R-3 occupancy located in a mixed occupancy building or within a Group R-3 or R-4 congregate living facility.

f. The length of common path of egress travel distance in a Group S-2 open parking garage shall be not more than 100 feet.

~~or ramps. Where access to exits occurs from adjacent building levels, the~~ ~~horizontal and vertical exit access travel distance to the closest exit shall~~ ~~not exceed that specified in Section 1016.1. Access to exits at other levels~~ ~~shall be from an adjacent story.~~ The path of egress travel to an *exit* shall not pass through more than one adjacent story.

Each story above the second story of a building shall have not less than one interior or exterior exit stairway, or interior or exterior exit ramp. Where three or more exits or access to exits are required, not less than 50 percent of the required exits shall be interior or exterior exit stairways or ramps.

Exceptions: ~~Landing platforms or roof areas for helistops that are~~ ~~less than 60 feet (18 288 mm) long, or less than 2,000 square feet~~ ~~(186 m²) in area, shall be permitted to access the second exit by a~~ ~~fire escape, alternating tread device or ladder leading to the story or~~ ~~level below.~~

1. Interior exit stairways and interior exit ramps are not required in open parking garages where the means of egress serves only the open parking garage.

2. Interior exit stairways and interior exit ramps are not required in outdoor facilities where all portions of the means of egress are essentially open to the outside.

~~**SECTION 1021**~~
~~**NUMBER OF EXITS AND EXIT CONFIGURATION**~~

(Relocated 1021.1 to 1021.4 to new 1006)

SECTION 1007
EXIT AND EXIT ACCESS DOORWAY CONFIGURATION

CHANGE SIGNIFICANCE: The way the code is laid out now, Section 1006 will provide information related to the number of means of egress that are required. This is done by Section 1006.2 addressing the egress from spaces, Section 1006.3 dealing with the number of egress routes from stories or occupied roofs, and Section 1007 addressing the arrangement or separation of those required means of egress.

A primary technical aspect to this change is that while the previous code included information within Sections 1014, 1015, and 1021 dealing with the number and arrangement of means of egress required, the 2015 code has now consolidated these requirements into Sections 1006 and 1007. Perhaps the most apparent change will be seen in the new Table 1006.2.1. The format for this new table is similar to previous Table 1021.2(2) and combines the occupant load requirements from the previous Table 1015.1 along with the common path of egress travel provisions from Section 1014.3 into a single location. Putting these two variables (design occupant load and occupant remoteness) into a single table should make determining the number of means of egress required from a space easier and lead to a more consistent application of the code.

The definition for "common path of egress travel" has also been changed so it more closely aligns with the definition for "exit access travel distance," but it terminates at an earlier point (where two egress routes

1006, 1007 continues

1006, 1007 continued

are available). The definition has been also revised to eliminate the sentence addressing paths that merge. Since the path is measured to the point where "occupants have separate and distinct access" to two exits or exit access doorways, if the paths do merge they would no longer provide the two distinct paths and therefore would still be considered within the common path of egress travel measurement.

Perhaps the biggest technical code change is found within Tables 1006.3.2(1) and 1006.3.2(2) where the last column has been changed from limiting the exit access travel distance to now limiting the common path of egress travel. While the intention based on the table's title heading is to limit stories that have only a single means of egress, the fact that the distance specified can be measured to the point where distinct and separate means of egress are available instead of to the door of an actual exit is a substantial change. For example, consider a small second floor with only one stairway from it that sits above a larger first floor area that has two exterior doors leading from it. Previously the occupants on the second floor were required to reach the protection of an exit within the specified travel distance shown in the table. Under the new provision, because it is now limiting the "common path of egress travel," it can allow a measurement that follows the occupants across the second floor, down an open exit access stairway and then stop that measurement once they have reached the first floor where they do finally have "separate and distinct access to two exits." Under the 2012 code they would have needed to reach the safety of an exit at this distance. Under the 2015 code they are simply given a second option for egress and could continue on across the unprotected first floor level for another 125 to 225 feet depending on the occupancy and whether the building is sprinklered. Therefore, while the 2012 code was more limiting by requiring the occupants of stories with a single means of egress to reach the safety of an exit within the specified distance, the new provisions will be less restrictive and not require the safety of an exit but instead just impose the second egress path.

Code users should also note a clarification that occurred within Section 1006.3 where the path of egress travel from one story through an adjacent level is limited to "one adjacent story." It has always been the intent of the code to limit an occupant's egress travel on an exit access stairway to travel only between adjacent stories. In the 2012 code, this requirement was the intent of the sentence in Section 1021.3.1 that stated: "Access to exits at other levels shall be from an adjacent story." The new language should help clarify the requirement and prohibit having an occupant travel more than one story via an exit access stairway or exit access ramp to reach an exit that is located on another story.

Section 1007 dealing with the separation or arrangement of the means of egress now provides specific guidance regarding how the distance between doors, exit access stairways, or exit access ramps is to be measured. Previously, the code did not provide specific measuring points for determining the exit or exit access separation and therefore it occasionally led to confusion or differences in application. With the new Section 1007.2, the measurements should be more consistent and less subjective.

CHANGE TYPE: Modification

CHANGE SUMMARY: This section now provides specific information regarding the point where exit separation is to be measured. Where three or more means of egress are required, the code restores performance language to ensure the egress paths are adequately separated.

2015 CODE: **1007.1.1.1 Measurement Point.**   The separation distance required in Section 1007.1.1 shall be measured in accordance with the following:

1. The separation distance to exit or exit access doorways shall be measured to any point along the width of the doorway.
2. The separation distance to exit access stairways shall be measured to the closest riser.
3. The separation distance to exit access ramps shall be measured to the start of the ramp run.

1007.1 continues

1007.1

Exit and Exit Access Doorway Configuration

Separation measured at any point along with width of the doorway

Additonal route "a reasonable distance apart" to prevent both being blocked

Separation as required by Section 1007.1.1 for two of the doorways:
• one-half of overall diagonal, or
• one-third of diagonal if sprinklered.

International Code Council®

Space requiring three or more means of egress

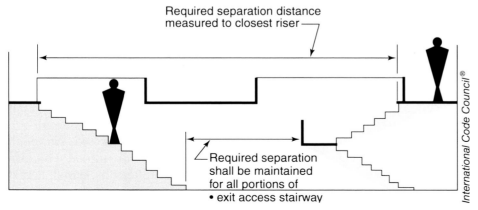

Required separation distance measured to closest riser

Required separation shall be maintained for all portions of
• exit access stairway

International Code Council®

Measurement point for exit access stairways

1007.1 continued

1007.1.2 ~~1015.2.2~~ Three or More Exits or Exit Access Doorways. Where access to three or more exits is required, ~~at least~~ <u>not less than</u> two exit or exit access doorways shall be arranged in accordance with the provisions of Section <u>1007.1.1</u> ~~1015.2.1~~. <u>Additional required exit or exit access doorways shall be arranged a reasonable distance apart so that if one becomes blocked, the others will be available.</u>

1007.1.3 Remoteness of Exit Access Stairways or Ramps. <u>Where two exit access stairways or ramps provide the required means of egress to exits at another story, the required separation distance shall be maintained for all portions of such exit access stairways or ramps.</u>

1007.1.3.1 Three or More Exit Access Stairways or Ramps. <u>Where more than two exit access stairways or ramps provide the required means of egress, not less than two shall be arranged in accordance with Section 1007.1.3.</u>

CHANGE SIGNIFICANCE: Section 1007.1.1.1 provides specific guidance on how the separation or remoteness of the means of egress is to be determined. This section provides three measurement methods that clearly state how the measurement between doors, stairways and ramps is to be done. This will provide clarity and consistency and will eliminate confusion or some of the inconsistent code application that has been common under the previous editions of the code. For example, under the previous code when an exit access stairway was used, the point where the path of travel entered the stairway was by definition an "exit access doorway." There was often concern as to whether the separation distance was measured there at the stair or at a point remote from it, perhaps the length of a required landing away. With this new section each item is specifically addressed, and it is now clear as to where the measurement is made.

This section also coordinates with new sections 1007.1.3 and 1007.1.3.1 so that the minimum separation distances between exit access stairways and ramps are maintained for the entire length of travel on the stairway or ramp. This is done to prohibit stairway and ramp runs that meet the separation distance requirement at the first riser or beginning slope from converging toward another stair or ramp such that the separation is reduced below the minimum distance as the occupant goes either up or down the stairway or ramp. It is reasonable to expect the egress separation distance to be maintained in order to ensure a fire cannot affect all of the egress paths, whether the separation distance is at the beginning of the stair or ramp or where the egress travel is completed. Code users will also notice that additional changes have been made within Section 1007.1.1 to make certain that the separation provisions apply to the doors, stairs, and ramps and not simply the exits or "exit access doorways" that were previously stated.

Where three or more means of egress are required, the code has added performance language back into Section 1007.1.2 in order to address the location of the third or fourth path of travel. The restored sentence was in the 2003 code but was removed from the 2006 edition because it was felt to be too subjective. After the sentence was removed, there was nothing to describe where the third or fourth element was to be located. Although this language is subjective, it at least provides the code official with language that can be used to accomplish the intent of the code, which is that a single fire is unlikely to take out multiple means of egress because they will be adequately separated and located remotely within the building. Code users may wish to look at code changes E127-12 and E7-12 to see how these two have been placed into the code. While the change in E7-12 would have applied this "reasonable distance apart" requirement to exit access stairways by virtue of the "exit access doorways" language found in the 2012 code, code change E127-12's addition of Sections 1007.1.3 and 1007.1.3.1 did not transfer the requirement to the stairway or ramp provisions. It is this author's opinion that the third and fourth required exit access stairway or ramp should also "be arranged a reasonable distance apart so that if one becomes blocked, the others will be available." It seems reasonable that this language should have also been included within Section 1007.1.3.1 but unfortunately was not due to the code changes being processed separately.

1009.8

Two-Way Communication

CHANGE TYPE: Clarification

CHANGE SUMMARY: This change clarifies that a two-way communication system may serve multiple elevators and that the systems are not required at service elevators, freight elevators, or private residence elevators.

Two-way communication system may serve multiple elevators

2015 CODE: 1009.8 ~~1007.8~~ Two-Way Communication.   A two-way communication system <u>complying with Sections 1009.8.1 and 1009.8.2</u> shall be provided at the <u>landing serving each</u> elevator ~~landing~~ <u>or bank of elevators</u> on each *accessible* floor that is one or more stories above or below the *story* of *exit discharge* ~~complying with Sections 1007.8.1 and 1007.8.2~~.

Exceptions:

1. Two-way communication systems are not required at the <u>landing serving each</u> elevator ~~landing~~ <u>or bank of elevators</u> where the two-way communication system is provided within *areas of refuge* in accordance with Section ~~1007.6.3~~ <u>1009.6.5</u>.

2. Two-way communication systems are not required on floors provided with ramps conforming to the provisions of Section ~~1010~~ <u>1012</u>.

3. <u>Two-way communication systems are not required at the landings serving only service elevators that are not designated as part of the accessible means of egress or serve as part of the required accessible route into a facility.</u>

4. <u>Two-way communication systems are not required at the landings serving only freight elevators.</u>

5. <u>Two-way communication systems are not required at the landing serving a private residence elevator.</u>

CHANGE SIGNIFICANCE: These changes help clarify which elevator landings are required to have a two-way communication system. Perhaps the most helpful aspects of this change are the exceptions that address service elevators that are not used for an accessible route or as part of the accessible means or egress, and also freight elevators and private residence elevators. Based on the previous code language, the two-way communication system requirement was often applied inconsistently because while some jurisdictions did apply the provision only to the elevator landings serving passenger elevators, others applied it to each individual elevator and type of elevator, including freight elevators located in back-of-house areas.

The revised text indicates that the communication system is not required at each elevator but at the landing serving the elevators. This can allow a single communication system to be installed and serve a number of elevators that open at the same landing. For example, if an elevator lobby has perhaps six or eight elevators opening into it, a single two-way communication system within the lobby would be adequate to satisfy the requirement that a system is "provided at the landing serving each elevator or bank of elevators." The intent of the new wording is that whether there is a single elevator or multiple elevators opening onto the landing, a single two-way communication system would be acceptable.

The new exceptions address conditions where an elevator landing would not require the installation of a two-way communication system. Exception 4 exempts landings that serve only freight elevators. Because freight elevators are not intended for passenger use, there is no reason for the communication system at the landing serving them. If the landing

1009.8 continues

1009.8 continued

serves both freight and passenger elevators, then the exception is not applicable, and the two-way communication system is required. Because private-residence elevators are allowed only within individual dwelling units, Exception 5 eliminates the communication system requirements from the landings where this type of elevator is installed. Within an individual dwelling unit, a two-way communication system is not needed or practical.

Exception 3 contains a number of items that the code official must consider before he or she eliminates the communication system. If an elevator is intended only as a service elevator, then the communication system may not be necessary at the landing. This exception could be used for many back-of-house service elevator installations, provided the landing is only for the service elevator and not other passenger elevators and that this elevator is not designated as a part of the accessible means or egress or used at the accessible route to the floor. The key difference between Exception 3 and Exception 4 is that the service elevators are passenger-type elevators and therefore additional limits are applicable in order for them to comply with Exception 3. As mentioned previously, Exception 4 is intended only for freight elevators that are not intended for passenger use.

One item for the code official to consider when applying this code change is to focus on the fact the communication system is required "at the landing." It is best to ignore the wording about a "bank" of elevators and apply the requirement like it was saying the communication system is required "at the landing" serving a single elevator or multiple elevators. A "bank" of elevators is all of the cars that respond to a single call button. The intent of the code provision is to have a communication system at each landing or lobby area serving elevators regardless of the number of elevators or the number of "banks" or control systems involved. For example, if an elevator lobby has three elevators on one side that only go up and three elevators on the other side that only go down, the lobby is probably served by two separate "banks" of elevators, since each call system has only three elevators which respond to it. In this situation it is reasonable that a single two-way communication system is installed in the lobby even though it is served by two "banks" of elevators. Because the landing serves both banks of elevators, a single communication system would still allow any occupants at that level to notify someone of their location, and the complying communication system would be "provided at the landing" serving each bank of elevators, just as the code requires.

CHANGE TYPE: Modification

CHANGE SUMMARY: Numerous revisions throughout these locking provisions help clarify requirements and their application by using consistent terminology. These changes allow an existing locking system exception for main doors that are not located at the exterior of the building.

2015 CODE: Because this code change affected substantial portions of Section 1010.1.9, the entire code change text is too extensive to be included here. Refer to Code Changes E-62, E-63, E-66, E-67, E-69, E-70, E-72, E-73, E-74, E-77, E-78, E-80, E-81, E-82, E-9, and E-2 in the *2015 IBC Code Changes Resource Collection* for the complete text and history of the code change.

1010.1.9 ~~1008.1.9~~ **Door Operations.**   Except as specifically permitted by this section, egress doors shall be readily openable from the egress side without the use of a key or special knowledge or effort.

1010.1.9.3 ~~1008.1.9.3~~ **Locks and Latches.**   Locks and latches shall be permitted to prevent operation of doors where any of the following exists:

1. (No change)
2. In buildings in occupancy Group A having an occupant load of 300 or less, Groups B, F, M and S, and in places of religious worship, the main ~~exterior~~ door or doors are permitted to be equipped with key-operated locking devices from the egress side provided:
   2.1. The locking device is readily distinguishable as locked;
   2.2. A readily visible durable sign is posted on the egress side on or adjacent to the door stating: THIS DOOR TO REMAIN UNLOCKED WHEN ~~BUILDING~~ <u>THIS SPACE</u> IS OCCUPIED. The sign shall be in letters 1 inch (25 mm) high on a contrasting background; and
   2.3. The use of the key-operated locking device is revocable by the building official for due cause.

1010.1.9 continues

1010.1.9

Door Operations— Locking Systems

Electromagnetic lock

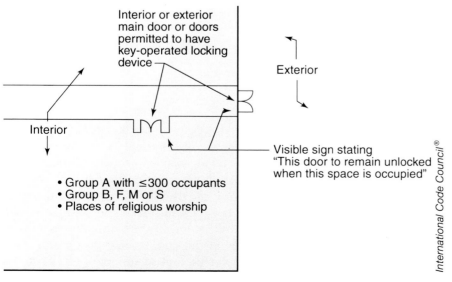

Interior or exterior main door or doors permitted to have key-operated locking device

Exterior

Interior

Visible sign stating "This door to remain unlocked when this space is occupied"

- Group A with ≤300 occupants
- Group B, F, M or S
- Places of religious worship

Key-locking hardware permitted on interior doors

1010.1.9 continued

3. (No change)

4. (No change)

5. (No change)

1010.1.9.6 ~~1008.1.9.6 Special~~ **Controlled** Egress ~~Locking Arrangements in~~ **Doors in Groups I-1 and** ~~Group~~ **I-2.** ~~Approved, special egress locks~~ Electric locking systems, including electro-mechanical lock-ing systems, and electromagnetic locking systems, shall be permitted to be locked in the means of egress in ~~a~~ Group I-1 or I-2 ~~occupancy~~ occupancies where the clinical needs of persons receiving care require their containment. ~~such locking. Special egress locks~~ Controlled egress doors shall be permitted in such occupancies where the building is equipped throughout with an automatic sprinkler system in accordance with Section 903.3.1.1 or an approved automatic-smoke or heat detection system installed in accordance with Section 907, provided that the doors are installed and operate in accordance with all of the following.

(Items 1 through 8 and two exceptions are not shown.)

1010.1.9.7 ~~1008.1.9.7~~ **Delayed Egress** ~~locks.~~

1010.1.9.8 ~~1008.1.9.8 Access Controlled~~ **Sensor Release of Electrically Locked Egress Doors.** The electric locks on sensor released ~~The entrance~~ doors located in a means of egress in buildings with an occupancy in Groups A, B, E, I-1, I-2, I-4, M, R-1 or R-2 and entrance doors to tenant spaces in occupancies in Groups A, B, E, I-1, I-2, I-4, M, R-1 or R-2 are permitted ~~to be equipped with an approved~~ entrance and egress access ~~control system, listed in accordance with UL 294, which shall be~~ where installed and operated in accordance with all of the following criteria:

(*Items 1 through 6 are not shown*)

1010.1.9.9 ~~1008.1.9.9~~ **Electromagnetically Locked Egress Doors.** Doors in the *means of egress* in buildings with an occupancy in Group A, B, E, I-1, I-2, I-4, M, R-1 or R-2 and doors to tenant spaces in Group A, B, E, I-1, I-2, I-4, M, R-1 or R-2 shall be permitted to be ~~electromagnetically~~ locked with an electromagnetic locking system where ~~if~~ equipped with ~~listed~~ hardware that incorporates a built-in switch ~~and meet the requirements below~~ and are installed and operated in accordance with all of the following:

(*Items 1 through 6 are not shown*)

CHANGE SIGNIFICANCE: The locking requirements contain a substantial number of changes that when looked at separately may not seem significant, but since there are so many changes, code users need to be aware of them and review the provisions so they are properly applied. Many of the changes have been made to coordinate the language within the code with the terminology used within the lock industry and within the UL 294 standard, which is referenced by several of the sections. Using consistent terminology helps clarify what the code is accepting, and it allows designers, manufacturers, inspectors, and installers to be asking for the same type of device or system. Some of the more quickly apparent changes include the terminology changes within the section titles such as "special locking arrangements" becoming "controlled egress doors in Groups I-1 and I-2" and "access controlled egress doors" becoming "sensor release of electrically locked egress doors."

The following is a brief overview of some of the changes and their intent:

Section 1010.1.9.3:   The provision that allowed the main exit door to be locked with key-locking hardware when a sign was posted adjacent to the door has been modified so that the provision is not limited to exterior doors but can instead be applied to the main egress door from an interior space within a building. This will allow the provision to apply to the door from a lecture hall within a college building or for something such as a movie theater or restaurant located within a covered mall building. As a part of this change, the language on the sign has also been revised so that it addresses the space served by that egress door and not the building.

Section 1010.1.9.6:   The title of the section and the terminology within it has been changed to be more specific and consistent with that used in the industry and to be more technically correct. Provisions have been expanded to include Group I-1 occupancies with the previously accepted Group I-2 occupancies, since the concern with elopement of the patients is also a concern for occupants of the Group I-1 uses. An exception has been added that eliminates requirements 1 through 4 of the provisions for areas such as hospital nurseries or obstetric areas where abduction of a child is a concern. This exception mirrors the previous exception that allowed patient treatment or containment areas to eliminate the automatic unlocking requirements and the limitation of passing through only one controlled door.

Section 1010.1.9.7:   Revisions include a number of terminology changes for consistency with the UL 294 standard as well as to clarify that these are delayed-egress locking systems and not just locks by themselves. Item 4 has been modified so the person attempting to egress must make a physical effort to open the hardware for "not more than 3 seconds." Previously the limit was 1 second, which is not enough time for cognizant persons to realize that their actions are causing the alarm and to back away prior to the system unlocking the door.

The limitation of not passing through more than one delayed egress door has been moved down into the list of numbered requirements. This change has been made to include an exception allowing occupants of Groups I-2 and I-3 to pass through a maximum of two doors, provided the overall delay does not exceed 30 seconds. Having multiple doors can help with preventing resident elopement, and yet the overall delay would still be equal or less than what the code has previously accepted for "approved" situations. An example where the two door arrangement may be helpful would be a multi-story facility where the door from the story and then also the door from the building could be controlled. A maximum 30-second delay in these very regulated, fully sprinklered facilities was viewed as being reasonable.

The signage requirements within Item 6 have been modified to recognize that the doors could open opposite the direction of travel and swing into the space. Therefore, the provision provides language for the sign,

1010.1.9 continues

1010.1.9 continued   which is dependent upon the direction of the door swing. Previously, the wording for the sign addressed only the doors swinging in the direction of egress travel. An exception has also been added to allow the signs to be eliminated where the patients require restraint or containment to prevent elopement. Based on the staff training within these facilities, the need to protect the residents by preventing elopement and the fact these systems are required to be interconnected with the sprinkler or fire detection system and unlock upon loss of power, the proponent and ICC membership believed that eliminating the sign in these facilities was reasonable.

Section 1010.1.9.8: The title and language within this section has been revised to point out it is a sensor release of a locked egress door and not the access to the space that is being controlled. Group I-1 and I-4 occupancies have been added to the list of acceptable uses due to the concern for security and safety and for patient elopement or an occupant leaving a day care facility. Since these types of locked egress doors allow for some control while at the same time allowing for free egress during an emergency, it seemed appropriate to include the I-1 and I-4 occupants where these needs could be addressed. In the numbered list of requirements, a provision was added as item 6, which will require these systems to comply with UL 294. The previous Item 6, which required the doors serving a Group A, B, E, or M occupancy to be open when the building was open to the public, has been eliminated because if the doors do comply with all of the other required items, the doors are ensured to be available for egress.

Section 1010.1.9.9: The requirements for electromagnetically locked egress doors have been changed to include the Group I-1, I-2, and I-4 occupancies for security and safety reasons, which were discussed with Section 1010.1.9.8. The important distinction here is that Group I-2 occupancies are added into this section for the 2015 code but were previously allowed in the provisions shown in Section 1010.1.9.8. Compliance with UL 294 has also been added. This helps ensure consistency between the various locking sections and helps ensure that these systems will perform as intended.

Section 1010.1.10: An exception has been added to the panic hardware requirements for Group A and E occupancies. This exception will allow electromagnetically locked systems provided they comply with Section 1010.1.9.9.

1011.15, 1011.16

Ship Ladders and Ladders

CHANGE TYPE: Addition

CHANGE SUMMARY: This section has been added to list the locations where ladders can be used for access. Permanent ladders must follow the construction requirements from the IMC in order to provide consistent installation and a safe usable ladder.

2015 CODE: **1011.15 ~~1009.14~~ Ships Ladders.** Ship ladders are permitted to be used in Group I-3 as a component of a means of egress to and from control rooms or elevated facility observation stations not more than 250 square feet (23 m²) with not more than three occupants and for access to unoccupied roofs. <u>The minimum clear width at and below the handrails shall be 20 inches (508 mm).</u>

1011.15.1 Handrails of Ship Ladders. <u>Handrails shall be provided on both sides of ship ladders.</u>

1011.15.2 Treads of Ship Ladders. Ship ladders shall have a minimum tread depth of 5 inches (127 mm). The tread shall be projected such that the total of the tread depth plus the nosing projection is not less than 8½ inches (216 mm). The maximum riser height shall be 9½ inches (241 mm).
  ~~Handrails shall be provided on both sides of ship ladders. The minimum clear width at and below the handrails shall be 20 inches (508 mm).~~

1011.16 Ladders. <u>Permanent ladders shall not serve as a part of the means of egress from occupied spaces within a building. Permanent ladders shall be permitted to provide access to the following areas:</u>

1. <u>Spaces frequented only by personnel for maintenance, repair or monitoring of equipment.</u>
2. <u>Nonoccupiable spaces accessed only by catwalks, crawl spaces, freight elevators or very narrow passageways.</u>

1011.15, 1011.16 continues

Ladder access is permitted to spaces used only by personnel for maintenance.

Ladder access is permitted to limited spaces.

1011.15, 1011.16 continued

3. Raised areas used primarily for purposes of security, life safety or fire safety including, but not limited to, observation galleries, prison guard towers, fire towers or lifeguard stands.

4. Elevated levels in Group U not open to the general public.

5. Non-occupied roofs that are not required to have stairway access in accordance with Section 1011.12.1.

6. Ladders shall be constructed in accordance with Section 306.5 of the *International Mechanical Code.*

CHANGE SIGNIFICANCE: There are two main parts to the revisions shown within Section 1011. The first part is simply a reformatting of the "stairway" provisions to better address the various situations where ladders, ship ladders, and alternating tread devices are allowed as alternatives to the generally required stairways. The second part of the changes is the addition of Section 1011.16, which will specifically address permanent ladders that may be used to provide access to certain spaces. It also provides details of how the ladders are to be constructed.

In some situations, the code has not previously been clear on whether a means of egress is required from certain spaces such as catwalks above ceilings, mechanical equipment areas, and service pits. These types of spaces are often found within buildings and occasionally face different requirements depending on whether they are viewed as being for limited service access or for some other purpose. The new section provides a list of five locations where the code will permit a permanent ladder to be installed for both access and egress from these infrequently occupied or limited access spaces. In addition, the code references the provisions of IMC Section 306.5, which applies when mechanical equipment is located in an elevated space or on a roof. This new Section 1011.16 therefore helps delineate when ladders can be used to access certain areas that are not typically considered to be occupied and thus do not need to be served by a means of egress.

The other important aspect of this code change is the fact that by referencing the provisions of IMC 306.5, the IFC provides the details and construction requirements for these permanent ladders. This will help to ensure that the ladders are safe and useable and will provide consistency for both the designer and code official. Without including this reference, the code does not specify how the ladders are to be constructed or what level of performance they are expected to provide. Although some jurisdictions had previously relied on the IMC requirements, there was nothing within the IFC or IBC that mandated those provisions where the ladders were not serving mechanical equipment.

1014.8
Handrail Projections

CHANGE TYPE: Modification

CHANGE SUMMARY: This section now provides guidance and enforceable language so the building official can determine when a pair of intermediate handrails begins to obstruct the required egress width of a stairway. This helps to clarify when the width of the stair must be increased due to the two intermediate handrails reducing the available egress width.

2015 CODE: ~~1012.8~~ **1014.8 Projections.** On ramps and on ramped aisles that are part of an accessible route, the clear width between handrails shall be 36 inches (914 mm) minimum. Projections into the required width of aisles, stairways and ramps at each side shall not exceed 4-½ inches (114 mm) at or below the handrail height. Projections into the required width shall not be limited above the minimum headroom height required in Section ~~1009.5~~ 1011.3. Projections due to intermediate handrails shall not constitute a reduction in the egress width. Where a pair of intermediate handrails are provided within the stairway width without a walking surface between the pair of intermediate handrails and the distance between the pair of intermediate handrails is greater than 6 inches (152 mm), the available egress width shall be reduced by the distance between the closest edges of each such intermediate pair of handrails that is greater than 6 inches (152 mm).

CHANGE SIGNIFICANCE: Although this change has added text that modified the existing provision, it should not result in any technical change in the way the provision is enforced. It simply provides both the designer and code official with clearer language on how to address a situation that is fairly common.

1014.8 continues

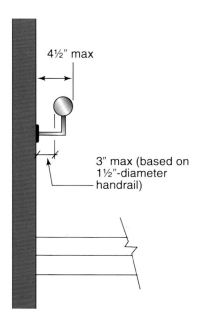

Typical handrail projection at wall

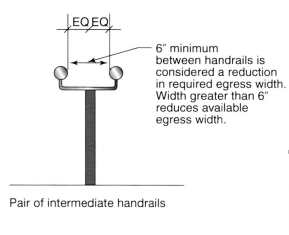

Pair of intermediate handrails

International Code Council®

1014.8 continued

Where an intermediate handrail is located on a stair or ramp, the code has previously stated that the intermediate handrails do not constitute a reduction in the egress width. Although this is easy enough to accept when there is only a single intermediate handrail, it becomes confusing if the intermediate handrail is actually constructed using double railings so that the users on each side have their own handrail to grasp as they egress. This code change describes the limitations for the spacing between the two handrails. Where the spacing between the two adjacent handrails is less than 6 inches, then the full width of the stairway can be used to satisfy the required egress width. If the space between the rails exceeds 6 inches, then the width of the stairway must be considered as being reduced by that excess amount.

To better understand the application of this provision, it is important to recognize that stairway width is typically measured at the walls located on the sides of the stairway. The code then allows for a handrail to project into that clear width up to 4½ inches. The code allows for this projection into the required width "at or below the handrail height." It is this 4½-inch allowed projection that serves as the basis for this provision. Assuming that the handrails are 1½ inches in diameter, the code would allow the inside edge of the handrail to be located 3 inches away from the wall. Allowing that 3-inch dimension on each side of an intermediate handrail is how the 6-inch dimension allowed by the code was determined. The figure helps to illustrate how the provision was developed and why it is not considered as creating a reduction in the egress width.

When applying this code provision, code officials will need to decide if they will follow the code's literal language of the 6-inch clearance or if they will follow the concept of how much of a projection is allowed on a stairway without creating a reduction in egress width. While the 6-inch clear space between the inside edges of the handrail works when the handrails are 1½ inches in diameter, the code actually allows handrails to have a circular cross section of "at least 1¼ inches and not greater than 2 inches" based on Section 1014.3.1. If a 1¼-inch handrail were used, the space between the wall and the inside edge of a handrail would be allowed to be 3¼ inches and therefore still within the 4½-inch maximum allowable projection limit. Conversely, if a 2-inch handrail were used, the clearance between the wall and the inside edge of the handrail would be limited to 2½ inches in order to still be within the 4½-inch limitation. Therefore, the gap between the inside edge of a pair of intermediate handrails could actually be as small as 5 inches or as large as 6½ inches, depending on the size of the handrail used if the intent is to limit the allowable projection to 4½ inches into each side. So if applied in a literal fashion, a 6-inch spacing requirement between a pair of 2-inch intermediate handrails would actually result in 5 inches of projection into each side of the stairway. Because of this discrepancy that can occur if handrails larger than 1½-inch sizing are used, code officials will need to decide if they follow the 6-inch limitation between the pair of intermediate handrails or the 4½-inch limit from the earlier part of the section. It is probably easiest for the code official to follow the language of the code and use the 6-inch dimension that is specified when an intermediate handrail consists of a pair of handrails with a gap between them.

Code users should also remember that this provision applies only to stairways and not to ramps. As stated earlier in the code text of this section, the clear width of a ramp is measured between the handrails, and projections are not allowed into that width.

CHANGE TYPE: Modification

CHANGE SUMMARY: This modification allows occupant egress through an elevator lobby provided access to at least one exit is available without the occupant passing through the lobby. It addresses the extent of the required elevator lobby protection.

2015 CODE: ~~1014.2~~ **1016.2 Egress through Intervening Spaces.** Egress through intervening spaces shall comply with this section.

> 1. <u>Exit access through an enclosed elevator lobby is permitted.</u>
>   <u>Access to at least one of the required exits shall be provided</u>
>   <u>without travel through the enclosed elevator lobbies required by</u>
>   <u>Section 3006.2, 3007 or 3008 of the *International Building Code.*</u>
>   <u>Where the path of exit access travel passes through an enclosed</u>
>   <u>elevator lobby, the level of protection required for the enclosed el-</u>
>   <u>evator lobby is not required to be</u>
>   <u>extended to the exit unless direct access to an exit is required by</u>
>   <u>other sections of this code.</u>

(No changes for previously existing four items and exceptions.)

~~1018.6~~ **1020.6 Corridor Continuity.**   Fire-resistance-rated corridors shall be continuous from the point of entry to an exit, and shall not be interrupted by intervening rooms. Where the path of egress travel within a

1016.2 continues

1016.2
Egress through Intervening Spaces

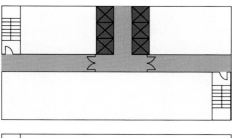
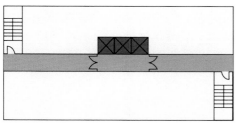
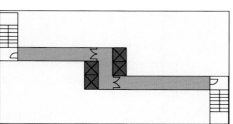

Exit access is permitted through an enclosed elevator lobby if access to at least one exit is provided without travel through the lobby.
Protection required for the lobby is not required to extend to an exit unless access to the exit is required by other sections (e.g., a fire service access elevator lobby requires direct access to an exit stairway).

International Code Council®

Egress through elevator lobby

1016.2 continued fire-resistance-rated corridor to the exit includes travel along unenclosed exit access stairways or ramps, the fire-resistance-rating shall be continuous for the length of the stairway or ramp and for the length of the connecting corridor on the adjacent floor leading to the exit.

Exceptions:

1. Foyers, lobbies or reception rooms constructed as required for corridors shall not be construed as intervening rooms.

2. Enclosed elevator lobbies as permitted by Item 1 of Section 1016.2 shall not be construed as intervening rooms.

IBC ~~713.14.1~~ 3006.4 Means of Egress. Elevator lobbies shall be provided with at least one means of egress complying with Chapter 10 and other provisions in this code. Egress through an elevator lobby shall be permitted in accordance with Item 1 of Section 1016.2.

IBC ~~3007.7~~ 3007.6 Fire Service Access Elevator Lobby. The fire service access elevator shall open into a fire service access elevator lobby in accordance with Sections 3007.6.1 through 3007.6.5. Egress is permitted through the elevator lobby in accordance with Item 1 of Section 1016.2.

Exception: Where a fire service access elevator has two entrances onto a floor, the second entrance shall be permitted to open into an elevator lobby in accordance with Section 708.14.1.

IBC ~~3008.7~~ 3008.6 Occupant Evacuation Elevator Lobby. The occupant evacuation elevators shall open into an elevator lobby in accordance with Sections 3008.6.1 through 3008.6.6. Egress is permitted through the elevator lobby in accordance with Item 1 of Section 1016.2.

CHANGE SIGNIFICANCE: This change helps resolve several questions that were not addressed or were unclear under the previous code. The section most affected is Section 1020.6, which contains the corridor-continuity provisions. Fire-resistance-rated corridors were required to be protected and to provide an egress path that was "continuous from the point of entry to an exit." These fire-resistance-rated corridors were not to be interrupted by intervening rooms; however, an exception did allow "foyers, lobbies, and reception areas" but did not clarify if an elevator lobby was intended to be included. With this change and with companion changes being made in the general elevator lobby provisions of IBC Section 3006.4 (previously Section 713.14.1) and the lobby requirements of IBC Sections 3007.6 and 3008.6 for fire service access or occupant evacuation elevators, the code will specifically allow and regulate how travel through the elevator lobby is to be addressed.

One of the key aspects of this change is that while egress is permitted through an enclosed elevator lobby, every occupant must have access to at least one exit without passing through the elevator lobby. This requirement ensures that if by chance smoke were to spread through the elevator hoistway and then out into the elevator lobby, at least one safe egress path is available to the occupants. Including this option within the code provides design flexibility without a reduction in safety but more importantly provides clarity, so the code is applied in a uniform manner.

This allowance for occupants passing through the elevator lobby as a part of one of the egress paths has been permitted in many parts of the country for years based on one of the legacy codes. There were no reports of this creating hazards and therefore it seemed best to address the issue directly within the code and to specify the limitations.

1017.2.2

Travel Distance Increase for Group F-1 and S-1 Occupancies

CHANGE TYPE: Modification

CHANGE SUMMARY: This modification allows an increased exit access travel distance within Group F-1 or S-1 occupancies meeting specific requirements. Also, it restores a travel distance that was allowed in the 2006 code but not allowed in the 2009 or 2012 editions.

2015 CODE: ~~1016.2.2~~ **1017.2.2 Group F-1 and S-1 Increase.**   The maximum exit access travel distance shall be 400 feet (122 m) in Group F-1 or S-1 occupancies where all of the following conditions are met:

1. The portion of the building classified as Group F-1 or S-1 is limited to one story in height.
2. The minimum height from the finished floor to the bottom of the ceiling or roof slab or deck is 24 feet (7315 mm).
3. The building is equipped throughout with an automatic fire sprinkler system in accordance with Section 903.3.1.1.

CHANGE SIGNIFICANCE: Group F-1 and S-1 occupancies are now allowed to increase their exit access travel distance from either 200 or 250 feet as allowed by Table 1017.2 up to a maximum of 400 feet when the two additional requirements from Section 1017.2.2 are imposed. The added requirements for the increased distance are that the F-1 or S-1 portion of the building is limited to a single story in height and that the space have a minimum height of 24 feet measured to either the ceiling or the roof deck. While there is a third requirement in Section 1017.2, that same requirement (that the building is sprinklered throughout in accordance with Section 903.3.1.1) would be applicable if the 250-foot exit-access travel distance from Table 1017.2 were to be used. Therefore, with

General requirement:
Without sprinkler system

200 feet →

With sprinkler system

250 feet →

Allowed by Section 1017.2.2 where

• area using increase is limited to single story in height, and
• minimum height to ceiling or roof is 24 feet, and
• building is sprinklered throughout.

400 feet →

Exit access travel distance for an F-1 or S-1 occupancy

International Code Council®

TABLE ~~1016.2~~ 1017.2   Exit Access Travel Distance[a]

Occupancy	Without Sprinkler System (Feet)	With Sprinkler System (Feet)
A, E, F-1, M, R, S-1	200	250[b]
F-2, S-2, U	300	400[c]

For SI: 1 foot = 304.8 mm.

a. See the following sections for modifications to exit access travel distance requirements:
   Section 412.7: For the distance limitations in aircraft manufacturing facilities.
   Section 1017.2.2: For increased distance limitation in Groups F-1 and S-1.

b. Buildings equipped throughout with an automatic sprinkler system in accordance with Section 903.3.1.1 or 903.3.1.2. See Section 903 for occupancies where automatic sprinkler systems are permitted in accordance with Section 903.3.1.2.

c. Buildings equipped throughout with an automatic sprinkler system in accordance with Section 903.3.1.1.

d. Group H occupancies equipped throughout with an automatic sprinkler system in accordance with Section 903.2.5.1.

(Only applicable portions of the table are shown.)

the two additional requirements shown in Section 1017.2.2 (Items 1 and 2), the maximum travel distance is allowed to go from 250 feet to 400 feet.

This 400-foot travel distance was previously allowed in the 2006 and earlier editions of the code, but it was tied to the requirement for the space to have smoke and heat venting in accordance with Section 910. The 400-foot travel distance option was removed from both the 2009 and 2012 code because thermally-activated vents were judged not to warrant such an increase. The reduction from 400 to 250 feet dramatically affected the layout and design of large storage or factory buildings due to the need for the designer to either change the proportions of the building or provide additional exits.

The 2015 code restores the 400-foot travel distance limit, which seems reasonable since the 400-foot distance has existed in the code and legacy codes for warehouses and factories with noncombustible materials since the early 1960s, and it was also allowed for all warehouses and factories for over a decade without any significant reports of problems. However, the reinstated travel distance is not based on the installation of smoke and heat vents but is instead based on fire modeling and egress times. The study used to support this change was commissioned and published by the California State Fire Marshal's Office. The "Report to the California State Fire Marshal on Exit Access Travel Distance of 400 Feet by Task Group 400, December 20, 2010" and the subsequent "Fire Modeling Analysis Report" dated July 20, 2011, provide the technical basis for increasing the travel distance and the required criteria. These reports can be found at http://osfm.fire.ca.gov/codedevelopment/pdf/2010interimcodeadoption/Part-9_ISOR_Attachment_A_20101221.pdf

The two criteria for the increased travel distance are that the increase is applicable only to portions of the building that are one story in height and that a 24-foot minimum ceiling height is provided. The first provision is intended to limit the area utilizing the 400-foot travel distance to the single-story requirement. Based on the wording about "the portion of the building," it would be permissible for the building to have some areas, such as offices, which were multistory, provided they use the general travel distance limits and that the area utilizing the 400-foot distance was

1017.2.2 continues

1017.2.2 continued

kept at the one-story limit. The ceiling or roof-deck height shown in Item 2 is based on the "Fire Modeling Analysis Report" and is used to provide a volume for the smoke to accumulate during the fire and provide additional time for egress.

Code users should be aware of two other travel distance changes within Table 1017.2. Footnote "a" has had an additional reference inserted that directs the user to Section 412.7 of the International Building Code (IBC) for aircraft manufacturing facilities. Section 412.7 has been added into the IBC and provides increased travel distance limits for aircraft manufacturing buildings of Type I or II construction based on both the height and area of the space. These travel distances can range from 400 to 1,500 feet depending on the building's size. These new requirements help address this unique limited-hazard occupancy and the large spaces they can require. Footnote "d" has also been added to address the sprinkler requirements for Group H occupancies. Since Group H occupancies are required to be sprinklered only within the occupancy and not throughout the building, the footnote has been added to distinguish the requirement from that of footnote "c." The code official should consider this requirement in conjunction with the limitations of Section 1016.2 Item 2 and its exception, which prohibits the occupants of a Group H occupancy from passing through a more hazardous use. Given this restriction, the travel distance limitations were viewed as being acceptable whether the travel was all within the Group H occupancy and sprinklered or if the occupants ended up leaving the Group H and egressing through another nonsprinklered occupancy.

CHANGE TYPE: Modification

CHANGE SUMMARY: The required width of aisles in Groups B and M occupancies as well as aisles in other occupancies are now tied to the widths required for corridors and not just to the capacity based on the occupant load served.

1018.3, 1018.5
Aisles

1018.3, 1018.5 continues

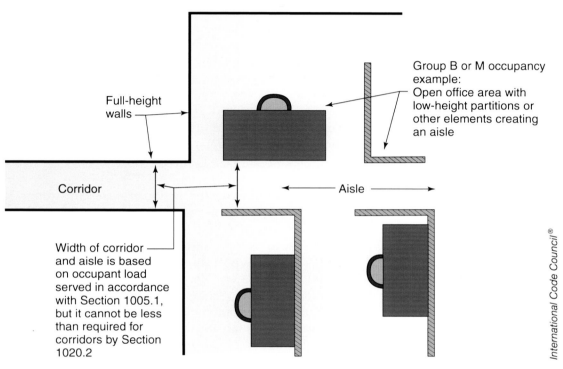

Full-height walls

Corridor

Aisle

Group B or M occupancy example:
Open office area with low-height partitions or other elements creating an aisle

Width of corridor and aisle is based on occupant load served in accordance with Section 1005.1, but it cannot be less than required for corridors by Section 1020.2

International Code Council®

Minimum aisle widths in Group B and M occupancies

International Code Council®

Group B occupancy

1018.3, 1018.5 continued

2015 CODE: ~~1017.3~~ <u>1018.3</u> **Aisles in Groups B and M.** In Group B and M occupancies, the minimum clear aisle width shall be determined by Section 1005.1 for the occupant load served, but shall ~~not~~ be <u>not</u> less than ~~36 inches (914 mm)~~ <u>that required for corridors by Section 1020.2</u>.

> **Exception:** Nonpublic aisles serving less than 50 people and not required to be accessible by Chapter 11 need not exceed 28 inches (711 mm) in width.

~~**1017.5**~~ **1018.5 Aisles in other than Assembly Spaces and Groups B and M.** In other than rooms or spaces used for assembly purposes and Group B and M occupancies, the minimum clear aisle ~~width~~ <u>capacity</u> shall be determined by Section 1005.1 for the occupant load served, but <u>the width</u> shall <u>be not less than that required for corridors by Section 1020.2</u> ~~not be less than 36 inches (914 mm)~~.

> **Exception:** <u>Nonpublic aisles serving less than 50 people and not required to be accessible by Chapter 11 of the *International Building Code* need not exceed 28 inches (711 mm) in width.</u>

CHANGE SIGNIFICANCE: Previously the width of aisles in Groups B and M occupancies was based solely on the capacity required for the occupant load served and a minimum width of 36 inches. A similar requirement also applied to aisles in occupancies "other than assembly spaces and Groups B and M." This new provision makes the aisles consistent with the width required for a corridor. Therefore, the aisles will now have not only the occupant load capacity (Section 1005.3) as a requirement but also the minimum width for the component based on the references to Section 1005.1 (and therefore Section 1005.2) and to Section 1020.2.

Perhaps the easiest illustration of where this will create a change is an aisle that serves more than 50 occupants. Previously, a code official examining this aisle would have used the occupant load multiplied by the egress capacity factor (which is generally 0.2 inches per occupant) to determine the required aisle width. As an example, assuming an occupant load of 100 people, the minimum aisle width under the 2012 code would have been 36 inches, whereas under the 2015 code this aisle would need to be a minimum of 44 inches in width (100 occupants × 0.2 inches per occupant = 20 inches; therefore, the 36 inch minimum was required).

Aisles are the main path of egress through many types of spaces such as cubicles in an open office area and between the merchandise pads in a display area of a store. While not always confined by the walls as a corridor is, it is reasonable that they should still be sized consistently with corridors so that the occupants could exit the building safely. Having the aisle width coordinated with the corridor requirements ensures the occupants will have a consistent egress width as they move from corridors to open spaces and vice versa, as might be found in an office building with both open office areas and other traditional support spaces.

The exception previously found within the code for nonpublic aisles in Groups B and M occupancies has been duplicated and placed in the section addressing other occupancies. Where the aisle serves a limited occupant load and is restricted, it is reasonable for the code official to allow the width to be reduced. Including the exception within Section 1018.5 provides consistency between the various non-assembly use groups.

CHANGE TYPE: Clarification

CHANGE SUMMARY: A new exception helps to clarify the width requirements for corridors within Group I-2 occupancies for areas where bed or stretcher movement is not necessary.

1020.2 continues

1020.2
Corridor Width and Capacity

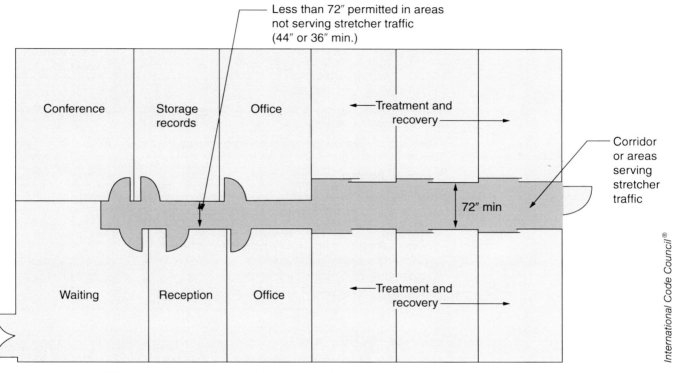

Reduced corridor width in areas not serving stretcher traffic or bed movement

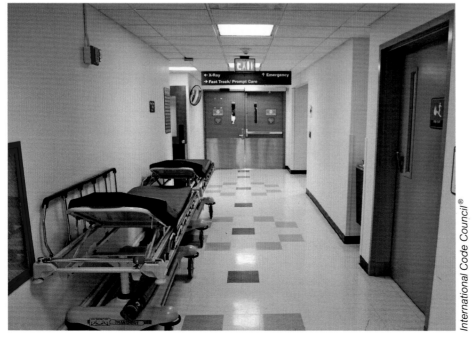

Bed movement in corridors requires greater corridor widths.

1020.2 continued

2015 CODE: ~~1018.2~~ 1020.2 **Width and Capacity.** The ~~minimum width~~ <u>required capacity</u> of corridors <u>shall be determined as specified in Section 1005.1, but the minimum width shall</u> ~~not~~ <u>be</u> not less than that specified in Table ~~1018.2~~ 1020.2 ~~shall be as determined in Section 1005.1~~.

> **Exception:** <u>In Group I-2 occupancies, corridors are not required to have a clear width of 96 inches (2438 mm) in areas where there will not be stretcher or bed movement for access to care or as part of the defend-in-place strategy.</u>

TABLE ~~1018.2~~ 1020.2   **Minimum Corridor Width**

Occupancy	Minimum width (inches)
Any facilities not listed below	44
Access to and utilization of mechanical, plumbing or electrical systems or equipment	24
With a<u>n</u> ~~required occupancy capacity~~ <u>occupant</u> <u>load of</u> less than 50	36
Within a dwelling unit	36
In Group E with a corridor having a<u>n</u> ~~required capacity~~ <u>occupant load</u> of 100 or more	72
In corridors and areas serving ~~gurney~~ <u>stretcher</u> traffic in ~~occupancies where patients receive outpatient medical care, which causes the patient to be incapable of self-preservation~~ <u>ambulatory care facilities</u>	72
Group I-2 in areas where required for bed movement	96

CHANGE SIGNIFICANCE: Since hospitals typically include accessory spaces or possibly non-separated mixed-use occupancies that are not used for patient care, the code official should have the clear ability to apply judgment in determining the appropriate means of egress requirements. For example, a large assembly space may need certain Group A requirements, while service and mechanical spaces with no patient care within them would not need an 8-foot corridor. Technically this option always existed in the previous code because the text of Table 1020.2 stated the wider width was needed in areas of Group I-2 "where required for bed movement." Therefore, the intent under the previous code would have been the same. The advantage of the new exception is that it specifically provides wording and guidance for the code official. It also serves as a reminder that consideration should be given not only to general patient care areas, but also to areas that may provide access to the care areas or that may be designated as patient relocation or refuge areas under the hospital's emergency plans. Therefore, if an area that may not normally be used for patient bed movement is designated as a part of the refuge area required by Section 407.5.1, then the wider aisle may still be needed within that space if there will be bed movement in that space.

The intent of the of the wider corridor is to allow for bed movement in opposite directions similar to the 44-inch general corridor provisions allowing people to pass in opposite directions. Therefore, where patients may be moved on beds, this wider corridor is needed. In other areas, the 44- or 36-inch-minimum width corridors could be acceptable if they adequately provide the required egress capacity.

This change has been another part of the ICC Ad Hoc Committee for Healthcare (AHC) efforts to evaluate, assess, and coordinate code issues related to hospitals and ambulatory healthcare facilities. This process is part of a joint effort between ICC and the American Society for Healthcare Engineering (ASHE), a subsidiary of the American Hospital Association.

1023.3.1
Stairway Extension

CHANGE TYPE: Modification

CHANGE SUMMARY: An interior exit stairway is now permitted to continue directly into an exit passageway without the need for a fire door assembly to separate the two elements.

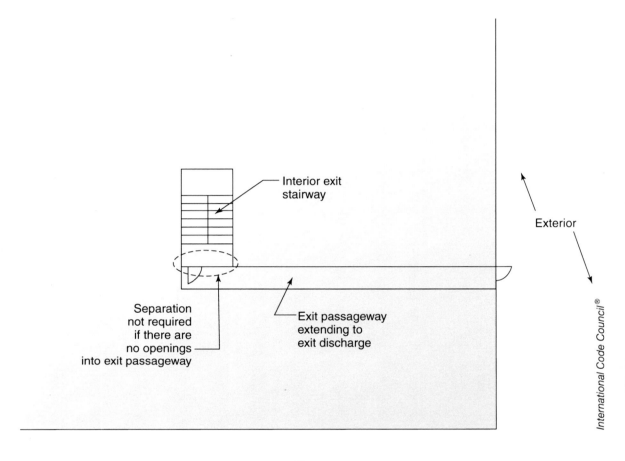

Interior exit
stairway

Separation
not required
if there are
no openings
into exit passageway

Exit passageway
extending to
exit discharge

Exterior

International Code Council®

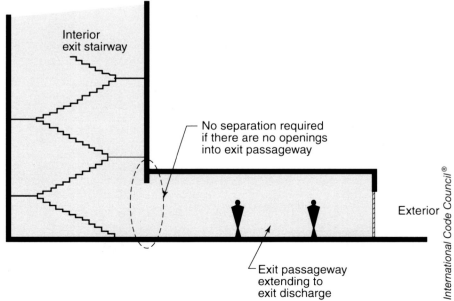

Interior
exit stairway

No separation required
if there are no openings
into exit passageway

Exit passageway
extending to
exit discharge

Exterior

International Code Council®

Separation between stairway and exit passageway can be eliminated

2015 CODE: ~~1022.3.1~~ **1023.3.1 Extension.** Where interior exit stairways and ramps are extended to an exit discharge or a public way by an exit passageway, the interior exit stairway and ramp shall be separated from the exit passageway by a fire barrier constructed in accordance with Section 707 of the *International Building Code* or a horizontal assembly constructed in accordance with Section 711 of the *International Building Code*, or both. The fire-resistance rating shall be ~~at least equal to~~ not less than that required for the interior exit stairway and ramp. A fire door assembly complying with Section 716.5 of the *International Building Code* shall be installed in the fire barrier to provide a means of egress from the interior exit stairway and ramp to the exit passageway. Openings in the fire barrier other than the fire door assembly are prohibited. Penetrations of the fire barrier are prohibited.

Exceptions:

1. Penetrations of the fire barrier in accordance with Section ~~1022.5~~ 1023.5 shall be permitted.

2. Separation between an interior exit stairway or ramp and the exit passageway extension shall not be required where there are no openings into the exit passageway extension.

CHANGE SIGNIFICANCE: Exit passageways are often used to extend an interior exit stairway from a remote location within the building to the exterior of the building. Besides making the exit continuous, this allows for the occupants to remain within the protected enclosure, and it maintains the required level of exit protection as is required by Section 1022.1. The base requirements of Section 1023.3.1 require that the interior exit stairway or ramp be separated from the exit passageway by a fire barrier and fire door assembly. The new exception will allow the separation between the horizontal and vertical exit elements to be eliminated when there are no openings into the exit passageway.

The general purpose of this door in the base requirement is to prevent smoke from an open door within the passageway from traveling up the vertical exit enclosure to other levels of the building. Where the new exception is followed and there are no openings into the passageway, there is no concern with smoke or fire entering the passageway and then compromising the exit enclosure. In addition, the egress travel can proceed quicker since there is no intermediate door through which the occupants must travel as they transition from the vertical exit enclosure to the horizontal travel within the exit passageway.

When applying this new exception it is important for the code official to recognize that this specific requirement mandates there are no openings into the exit passageway extension other than the connection from the interior exit stairway or ramp and the door providing egress from the passageway. This makes the exception more restrictive than the general exit passageway requirements found in Section 1024.5, which would typically allow doors providing egress from normally occupied spaces to open into the passageway. Penetrations into the passageway are not restricted by the new exception but would be limited as specified in Section 1024.6 to those elements serving the exit passageway itself. The same should also be applied to the ventilation of the exit passageway; it should be regulated and limited as addressed in Section 1024.7.

1023.3.1 continues

1023.3.1 continued

In many ways, it has always been questionable as to why the code would require a separation between the vertical exit enclosure and the exit passageway that extends it to the exit discharge given that the construction requirements are essentially equivalent. When the requirements of Sections 1022.1, 1023, and 1024 are applied, the construction requirements as well as the limitations on the types of openings, penetrations and ventilation result in an equivalent level of protection. Therefore, it does seem as if the door at the interface between the passageway and the vertical exit enclosure is needed any more than a door within the stair enclosure itself would be needed. Perhaps the only situation where the added restrictions from Section 1023.3.1 can truly be justified is where the passageway is improperly constructed or if it is within a mall building where the provisions of Sections 1024.5 and 402.8.7 do allow service areas to open into an exit passageway. Otherwise, the exit passageway provides the same level of protection for the egress path and therefore would seemingly be no more hazardous than a transition within the stairway or from an exit stair to an exit ramp would be.

1029.13.2.2.1

Stepped Aisle Construction Tolerances

CHANGE TYPE: Modification

CHANGE SUMMARY: This new section limits the variation allowed between adjacent risers within a stepped aisle. The previous code did not limit the variation for these risers.

2015 CODE: ~~1028.11.2~~ **1029.13.2.2 Risers.**   Where the gradient of ~~aisle stairs~~ stepped aisles is to be the same as the gradient of adjoining seating areas, the riser height shall ~~not~~ be not less than 4 inches (102 mm) nor more than 8 inches (203 mm) and shall be uniform within each flight.

Exceptions:

1. Riser height nonuniformity shall be limited to the extent necessitated by changes in the gradient of the adjoining seating area to maintain adequate sightlines. Where nonuniformities exceed 3/16 inch (4.8 mm) between adjacent risers, the exact location of such nonuniformities shall be indicated with a distinctive marking stripe on each tread at the nosing or leading edge adjacent to the nonuniform risers. Such stripe shall be ~~a minimum of~~ not less than 1 inch (25 mm), and ~~a maximum of~~ not more than 2 inches (51 mm), wide. The edge marking stripe shall be distinctively different from the contrasting marking stripe.

2. Riser heights not exceeding 9 inches (229 mm) shall be permitted where they are necessitated by the slope of the adjacent seating areas to maintain sightlines.

1029.13.2.2.1 Construction Tolerances.   The tolerance between adjacent risers on a stepped aisle that were designed to be equal height shall not exceed 3/16 inch (4.8 mm). Where the stepped aisle is designed in accordance with Exception 1 of Section 1029.13.2.2, the stepped aisle shall be constructed so that each riser of unequal height, determined in the direction of descent, is not more than 3/8 inch (10 mm) in height

1029.13.2.2.1 continues

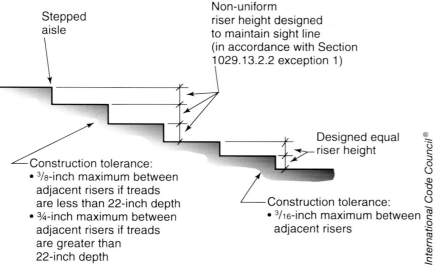

Stepped aisle

Non-uniform riser height designed to maintain sight line (in accordance with Section 1029.13.2.2 exception 1)

Designed equal riser height

Construction tolerance:
- 3/8-inch maximum between adjacent risers if treads are less than 22-inch depth
- 3/4-inch maximum between adjacent risers if treads are greater than 22-inch depth

Construction tolerance:
- 3/16-inch maximum between adjacent risers

International Code Council®

Limitations on riser height variations

1029.13.2.2.1 continued

different from adjacent risers where stepped aisle treads are less than 22 inches (560 mm) in depth and ¾ inch (19 mm) in height different from adjacent risers where stepped aisle treads are 22 inches (560 mm) or greater in depth.

CHANGE SIGNIFICANCE: Although construction tolerances for stairways are specified in Section 1011.5.4, they are not applicable to the stepped aisles covered by Section 1029. While the riser heights within a stepped aisle are allowed by Exception 1 in Section 1029.13.2.2 to vary "to the extent necessitated by changes in the gradient of the adjoining seating area to maintain adequate sightlines," the code never limited the actual difference between adjacent risers caused by variations in the construction. This new section will now limit the variation between risers that are intended to be equal in height as well as those that are intended to vary to maintain sightlines.

This change will effectively place limits on the variation that is allowed by Exception 1 of Section 1029.13.2.2. Previously because there were no limitations, the exception would allow any amount of variation between adjacent risers provided a distinctive marking stripe was placed at the nosing where variations exceeded $^{3}/_{16}$ inch. This basically allowed the exception to help support the lack of construction quality or diligence by permitting the striping to be used where risers that were designed to be equal in height varied by more than $^{3}/_{16}$ inch when they were actually constructed.

Under the new limitations in Section 1029.13.2.2.1, risers that "were designed to be equal height" will be limited to having a $^{3}/_{16}$-inch construction tolerance similar to that allowed for stairways in Section 1011.5.4. Where the stepped aisle risers are varied using Exception 1 in Section 1029.13.2.2 "to maintain adequate sightlines," the difference between adjacent risers is limited to either $^{3}/_{8}$ or ¾ inch depending on the tread depth. Previously the code would not have placed any restriction on the height change between adjacent risers other than the wording about being "limited to the extent necessitated . . . to maintain adequate sightlines." So while riser heights previously could have varied by an inch or more if that variation coincided with the gradient of the adjacent seating area, they will now have a maximum limit and be held to the $^{3}/_{8}$- or ¾-inch variation.

Having the construction tolerance and the maximum allowable designed variation limited will help to improve the safety of occupants using the stepped aisles for access to seating or for egress. It will also help ensure that the distinctive marking stripe required by Exception 1 in Section 1029.13.2.2 is more useful and limited to those situations where the riser heights are designed to vary due to the sightlines.

1103.4.1
Vertical Openings in Existing Group I-2 and I-3 Occupancies

CHANGE TYPE: Modification

CHANGE SUMMARY: Retroactive construction of a 1-hour fire-resistance-rated separation is now required in existing hospitals and jails to protect vertical openings. Alternatives have been included that can be used in lieu of the separation to mitigate the hazard created by the vertical openings.

2015 CODE: 1103.4.1 Group I I-2 and I-3 Occupancies. In Group I I-2 and I-3 occupancies, interior vertical openings connecting two or more stories shall be protected with 1-hour fire-resistance-rated construction.

Exceptions:

1. In Group I-2, unenclosed vertical openings not exceeding two connected stories and not concealed within the building construction shall be permitted as follows:

    1.1. The unenclosed vertical openings shall be separated from other unenclosed vertical openings serving other floors by a smoke barrier.

    1.2. The unenclosed vertical openings shall be separated from corridors by smoke partitions.

    1.3. The unenclosed vertical openings shall be separated from other fire or smoke compartments on the same floors by a smoke barrier.

    1.4. On other than the lowest level, the unenclosed vertical openings shall not serve as a required means of egress.

1103.4.1 continues

Vertical openings connecting two floors in existing Group I-2 occupancies are allowed to continue provided a minimum level of protection is provided.

1103.4.1 continued

2. In Group I-2, atriums connecting three or more stories shall not require 1-hour fire-resistance-rated construction when the building is equipped throughout with an automatic sprinkler system installed in accordance with Section 903.3, and all of the following conditions are met:

  2.1. For other than existing approved atriums with a smoke control system, where the atrium was constructed and is maintained in accordance with the code in effect at the time the atrium was created, the atrium shall have a smoke control system that is in compliance with Section 909.

  2.2. Glass walls forming a smoke partition or a glass-block wall assembly shall be permitted when in compliance with 2.2.1 or 2.2.2:

    2.2.1. Glass walls forming a smoke partition shall be permitted where all of the following conditions are met:

      2.2.1.1. Automatic sprinklers are provided along both sides of the separation wall and doors, or on the room side only if there is not a walkway or occupied space on the atrium side.

      2.2.1.2. The sprinklers shall not be more than 12 inches (304.8 mm) away from the face of the glass and at intervals along the glass of not greater than 72 inches (1829 mm).

      2.2.1.3. Windows in the glass wall shall be non-operating type.

      2.2.1.4. The glass wall and windows shall be installed in a gasketed frame in a manner that the framing system deflects without breaking (loading) the glass before the sprinkler system operates.

      2.2.1.5. The sprinkler system shall be designed so that the entire surface of the glass is wet upon activation of the sprinkler system without obstruction.

    2.2.2. A fire barrier is not required where a glass-block wall assembly complying with Section 2110 of the *International Building Code* and having a ¾-hour fire protection rating is provided.

  2.3. Where doors are provided in the glass wall, they shall be either self-closing or automatic-closing and shall be constructed to resist the passage of smoke.

3. In Group I-3 occupancies, exit stairways or ramps and *exit access stairways* or *ramps* constructed in accordance with Section 408 of the *International Building Code*.

CHANGE SIGNIFICANCE: The 2012 IFC contains a retroactive requirement to protect vertical openings connecting two stories or more in all Group I occupancies. This revised section is now limited to only Groups I-2 and I-3. By default, Groups I-1 and I-4 can now have unenclosed vertical openings connecting two stories and fall under the requirements for all other occupancies when they connect three or more stories. This is consistent with requirements for new construction, which now allow the use of open exit access stairways in all occupancies other than Groups I-2 and I-3.

This change makes the IFC consistent with federal standards that are in place for the protection of vertical openings in existing Group I-2 hospitals.

There are three exceptions to the requirement that all vertical openings be protected with fire-resistance-rated construction of at least 1 hour.

Exception 1 deals with Group I-2 occupancies with unenclosed vertical openings that do not connect more than two stories and are not concealed within the building construction, such as ventilation shafts or pipe chases. This would include designs such as an open stairway between two floors or a lobby open to the floor above. This is acceptable when the vertical opening is separate from other vertical openings and separated from other smoke compartments by smoke barriers. Additionally, the vertical opening must be separated from corridors by smoke partitions.

Exception 2 is specific to vertical openings created by an atrium in a Group I-2 occupancy. By definition, an atrium connects three or more stories, so where something looks like an atrium, but is only two stories, it will fall under Exception 1. Under Exception 2, atriums can continue as unenclosed vertical openings when they comply with all of the following items:

1. The building is protected with an automatic fire sprinkler system.

2. A smoke control system meeting the requirements at the time of construction is operational and maintained, or a smoke control system meeting the requirements of Section 909 is installed.

3. Where glass walls form a smoke partition, automatic fire sprinklers meeting certain spacing criteria must be provided on each side of the glass walls having an occupiable floor space.

4. Where a glass-block wall assembly is installed, it must meet IBC Section 2110 and have a ¾-hour fire-protection rating.

5. Doors in glass walls must be self-closing or automatic-closing and designed to resist the passage of smoke.

Exception 3 is specific to Group I-3 and allows unenclosed vertical openings that serve as exit stairways, exit ramps, exit access stairways, or exit access ramps provided they comply with IBC Section 408.

1103.7.6

Manual Fire Alarm Systems in Existing Group R-2 Occupancies

Manual fire alarm box

International Code Council®

CHANGE TYPE: Modification

CHANGE SUMMARY: The installation of interconnected smoke alarms within dwelling units along with fire-resistance-rated separation of dwelling units is allowed as an alternative to the retroactive installation of a manual fire alarm system throughout the building in existing Group R-2 occupancies.

2015 CODE: **1103.7.6 Group R-2.** A manual fire alarm system that activates the occupant notification system in accordance with Section 907.6 shall be installed in existing Group R-2 occupancies more than three stories in height or with more than 16 dwelling or sleeping units.

Exceptions:

1. Where each living unit is separated from other contiguous living units by fire barriers having a fire-resistance rating of not less than 0.75 hour, and where each living unit has either its own independent exit or its own independent stairway or ramp discharging at grade.

2. A separate fire alarm system is not required in buildings that are equipped throughout with an approved supervised automatic sprinkler system installed in accordance with Section 903.3.1.1 or 903.3.1.2 and having a local alarm to notify all occupants.

3. A fire alarm system is not required in buildings that do not have interior corridors serving dwelling units and are protected by an approved automatic sprinkler system installed in accordance with Section 903.3.1.1 or 903.3.1.2, provided that dwelling units either have a means of egress door opening directly to an exterior exit access that leads directly to the exits or are served by open-ended corridors designed in accordance with Section 1027.6, Exception 3.

4. A fire alarm system is not required in buildings that do not have interior corridors serving dwelling units, do not exceed three stories in height and comply with both of the following:

    4.1. Each dwelling unit is separated from other contiguous dwelling units by fire barriers having a fire-resistance rating of not less than ¾ hour.

    4.2. Each dwelling unit is provided with interconnected smoke alarms as required for new construction in Section 907.2.11.

CHANGE SIGNIFICANCE: The revision adds a new exception to the retroactive installation of a manual fire alarm system in existing Group R-2 occupancies. This section initially requires the retroactive installation of a manual fire alarm system in existing Group R-2 occupancies that do not have a fire alarm system. The new exception eliminates the requirement for a manual fire alarm system when the following four criteria are met:

1. The building does not exceed three stories in height,

2. The dwelling units are not served by interior corridors,

3. Each dwelling unit is separated by fire barriers with a minimum fire-resistance rating of ¾-hour, and

4. Interconnected smoke alarms are provided within each dwelling unit meeting the requirements for new construction.

The requirement for the installation of a manual fire alarm system is required only if the building exceeds three stories in height *or* contains more than 16 dwelling units. So, if the building contains more than 16 dwelling units but is less than three stories in height, this exception would apply if the additional criteria are met.

The issue of the building not having interior corridors results in building design that has either dwelling units with exits that lead directly to the outside or exterior egress balconies. In either design, the safety of the occupants is greatly enhanced since the egress balcony would be open to the outside air on one side and separated from the building on the other side. This is similar to the open-ended corridors referred to in Exception 3.

Each dwelling unit must be separated from contiguous dwelling units by walls or floor/ceiling assemblies with a ¾-hour fire-resistance rating. New construction would be separated by 1-hour fire-resistance-rated construction, so the allowance for a ¾-hour rating in existing buildings provides an acceptable level of mitigation. It is most likely that the existing construction is either non rated or 1-hour construction. To evaluate the equivalent fire-resistance value of the existing construction, Section 721 of the *International Building Code* can be utilized.

For example, consider a 2 x 4 wood-frame wall separating two dwelling units. Using the values in Section 722, the equivalent value of the existing construction can be calculated. This would give a time value of 20 minutes for wood studs at 16 inches on center, 15 minutes for ½-inch gypsum wallboard on one side of the wall assembly, and 15 minutes for various types of insulation in the wall cavity. Only the fire side of wall is considered when using these calculations, but this calculation would provide a fire-resistance rating of 50 minutes for that wall assembly, which would meet the criteria.

The most beneficial of the criteria is the requirement for interconnected smoke alarms within the dwelling unit. For years, the installation of a single smoke alarm within a dwelling unit was the requirement. This single smoke alarm would be placed in the common area between the sleeping rooms and the living area. The requirement for smoke alarms has evolved to the point that now smoke alarms are required on every floor level, in each sleeping room, and between the sleeping rooms and the living area, and they must be interconnected with a battery backup.

Retrofitting interconnected smoke alarms into all sleeping rooms in existing dwelling units provides an automatic level of safety to the dwelling-unit occupants as compared to a manual fire alarm system, which is waiting for someone to physically activate the manual fire alarm box.

1105

Construction Requirements for Existing Group I-2 Occupancies

CHANGE TYPE: Addition

CHANGE SUMMARY: Retroactive construction requirements have been added to the IFC to provide a minimum level for fire and life safety in existing Group I-2 occupancies.

2015 CODE:

SECTION 1105
CONSTRUCTION REQUIREMENTS FOR EXISTING GROUP I-2

1105.1 General.   Existing Group I-2 shall meet all of the following requirements:

1. The minimum fire safety requirements in Section 1103.
2. The minimum mean of egress requirements in Section 1104.
3. The additional egress and construction requirements in Section 1105.

Where the provisions of this chapter conflict with the construction requirements that applied at the time of construction, the most restrictive provision shall apply.

Note: An outline of the remainder of Section 1105 is shown here. For the entire text please refer to the 2015 IFC.

1105.2 Construction.
Table 1105.2—Floor Level Limitations for Group I-2 Condition 2
1105.3 Incidental Uses in Existing Group I-2.
Table 1105.3—Incidental Uses in Existing Group I-2 Occupancies
1105.3.1 Occupancy Classification.

Existing Group I-2 hospital

International Code Council®

CHANGE SIGNIFICANCE: This new section adds retroactive construction requirements for existing Group I-2 occupancies. These requirements align the IFC with the current approach by the Center for Medicaid and

1105 continues

1105 continued Medicare Services (CMS), the federal authority having jurisdiction. Hospitals are required by CMS to have a life safety survey on a regular basis. If the facility does not meet certain life safety minimums, it is required to upgrade its existing facility. The intent is to provide consistency between the two main regulatory agencies: the local jurisdiction and CMS.

These new requirements are a product of the ICC Ad Hoc Committee on Health Care. This committee consisted of code enforcement officials, regulatory agencies and the health care industry. All parties participated and commented on the requirements to ensure that these provisions were both necessary and achievable.

Some of the requirements in this section are

- Corridor walls must be designed to resist the passage of smoke (see Section 1105.4).
- Dutch doors (doors divided horizontally in such a fashion that the bottom half may remain shut while the top half opens) are allowed in corridor walls (see Section 1105.4.4.2.3).
- Mail slots are allowed in doors that are part of the corridor wall (see Section 1105.4.4.3).
- Protection of incidental use areas must be provided by either fire-resistance-rated separations or an automatic sprinkler system (see Table 1105.3).
- The minimum corridor width must be 48 inches (see Section 1105.5.5).
- The limitation of dead-end corridors is 30 feet (see Section 1105.5.6).
- Smoke compartments must be constructed (see Section 1105.6).
- A fire sprinkler system must be installed (see Section 1105.8).
- A fire alarm system must be installed (see Section 1105.9).

These provisions apply to Group I-2 occupancies. However, the Group I-2 occupancy must also comply with the general retroactive construction requirements in Sections 1103 and 1104 not modified or superseded by this new section.

PART 4

Special Occupancies

Chapters 20 through 49

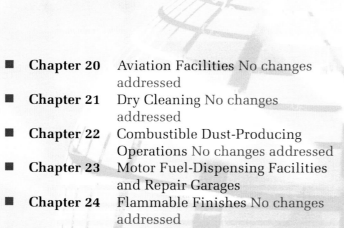
Chapters 20 through 36 of the *International Fire Code* (IFC) establish the minimum requirements for special processes and uses. These processes or uses involve a variety of combustible and hazardous materials and are unique to particular applications, such as semiconductor facilities or industrial ovens. For certain activities, such as aviation facilities or motor fuel-dispensing facilities, these processes or uses typically require the storage and handling of thousands of gallons of flammable or combustible liquids. Other chapters address the safe handling and storage of combustible and flammable materials. Chapters 37 through 49 are reserved and serve as placeholders for future code provisions addressing new special occupancies or operations that require regulation by the IFC. ■

2307.4

LP-gas Dispensing Operations

CHANGE TYPE: Modification

CHANGE SUMMARY: LP-gas requirements have been revised to improve correlation with other industry standards and to allow self-service LP-gas refueling by the public.

2015 CODE: 2307.4 Location of Dispensing Operations and Equipment. ~~In addition to the requirements of Section 2306.7, the point of transfer for LP-gas dispensing operations shall be 25 feet (7620 mm) or more from buildings having combustible exterior wall surfaces, buildings having noncombustible exterior wall surfaces that are not part of a 1-hour fire-resistance-rated assembly, or buildings having combustible overhangs, lot lines of property which could be built on, public streets, or sidewalks and railroads; and at least 10 feet (3048 mm) from driveways and buildings having noncombustible exterior wall surfaces that are part of a fire-resistance-rated assembly having a rating of 1 hour or more.~~ The point of transfer for LP-gas dispensing operations shall be separated from buildings and other exposures in accordance with the following:

1. Not less than 25 feet (7620 mm) from buildings where the exterior wall is not part of a fire-resistance-rated assembly having a rating of 1 hour or greater.

2. Not less than 25 feet (7620 mm) from combustible overhangs on buildings, measured from a vertical line dropped from the face of the overhang at a point nearest the point of transfer.

3. Not less than 25 feet (7620 mm) from the lot line of property that can be built upon.

4. Not less than 25 feet (7620 mm) from the centerline of the nearest mainline railroad track.

5. Not less than 10 feet (3048 mm) from public streets, highways, thoroughfares, sidewalks and driveways.

2307.4 continues

Self-service fueling operations now allow dispensing of LP-gas into motor vehicles by the public.

International Code Council®

2307.4 continued

6. Not less than 10 feet (3048 mm) from buildings where the exterior wall is part of a fire-resistance-rated assembly having a rating of 1 hour or greater.

Exception: The point of transfer for LP-gas dispensing operations need not be separated from canopies that are constructed in accordance with the *International Building Code* and that provide weather protection for the dispensing equipment.

LP-gas containers shall be located in accordance with Chapter 61. LP-gas storage and dispensing equipment shall be located outdoors ~~and in accordance with Section 2306.7~~.

2307.5 Additional Requirements for LP-gas Dispensers and Equipment. LP-gas dispensers and related equipment shall comply with the following provisions.

1. Pumps shall be fixed in place and shall be designed to allow control of the flow and to prevent leakage or accidental discharge.

2. Dispensing devices installed within 10 feet (3048 mm) of where vehicle traffic occurs shall be protected against physical damage by mounting on a concrete island 6 inches (152 mm) or more in height, or shall be protected in accordance with Section 312.

3. Dispensing devices shall be securely fastened to their mounting surface in accordance with the dispenser manufacturer's instructions.

~~2307.5~~ 2307.6 Installation of LP-gas Dispensing Devices and Equipment. The installation and operation of LP-gas dispensing systems shall be in accordance with Sections ~~2307.5.1~~ 2307.6.1 through ~~2307.5.3~~ 2307.6.4 and Chapter 61. LP-gas dispensers and dispensing stations shall be installed in accordance with the manufacturer's specifications and their listing.

~~2307.5.1 Valves.~~ ~~A manual shutoff valve and an excess flow-control check valve shall be located in the liquid line between the pump and the dispenser inlet where the dispensing device is installed at a remote location and is not part of a complete storage and dispensing unit mounted on a common base. An excess flow-control check valve or an emergency shutoff valve shall be installed in or on the dispenser at the point at which the dispenser hose is connected to the liquid piping. A differential backpressure valve shall be considered equivalent protection.~~

~~A listed shutoff valve shall be located at the discharge end of the transfer hose.~~

2307.6.1 Product Control Valves. The dispenser system piping shall be protected from uncontrolled discharge in accordance with the following:

1. Where mounted on a concrete base, a means shall be provided and installed within ½-inch of the top of the concrete base that will prevent flow from the supply piping in the event that the dispenser is displaced from its mounting.

2. A manual shutoff valve and an excess flow-control check valve shall be located in the liquid line between the pump and the dispenser inlet where the dispensing device is installed at a remote location and is not part of a complete storage and dispensing unit mounted on a common base.

3. An excess flow-control check valve or an emergency shutoff valve shall be installed in or on the dispenser at the point at which the dispenser hose is connected to the liquid piping.

4. A listed automatic-closing type hose nozzle valve with or without a latch-open device shall be provided on island-type dispensers.

~~2307.5.2~~ 2307.6.2 Hoses. Hoses and piping for the dispensing of LP-gas shall be provided with hydrostatic relief valves. The hose length shall not exceed 18 feet (5486 mm). An approved method shall be provided to protect the hose against mechanical damage.

2307.6.3 Breakaway Protection. Dispenser hoses shall be equipped with a listed emergency breakaway device designed to retain liquid on both sides of the breakaway point. Where hoses are attached to hose-retrieving mechanisms, the emergency breakaway device shall be located such that the breakaway device activates to protect the dispenser from being displaced.

~~2307.5.3~~ 2307.6.4 Vehicle Impact Protection. ~~Vehicle impact protection for LP-gas storage containers, pumps and dispensers shall be provided in accordance with Section 2306.4.~~ Where installed within 10 feet of vehicle traffic, LP-gas storage containers, pumps and dispensers shall be protected in accordance with Section 2307.5, Item 2.

2307.7 ~~Private~~ Public Fueling of Motor Vehicles. ~~Self-service LP-gas dispensing systems, including key, code and card lock dispensing systems, shall not be open to the public and shall be limited to the filling of permanently mounted fuel containers on LP-gas powered vehicles.~~ Self-service LP-gas dispensing systems, including key, code and card lock dispensing systems, shall be limited to the filling of containers providing fuel to the LP-gas powered vehicle.

~~In addition to the requirements of Sections 2305 and 2306.7,~~ The requirements for self-service LP-gas dispensing systems shall be in accordance with the following:

1. The arrangement and operation of the transfer of product into a vehicle shall be in accordance with this section and Chapter 61.

2. The system shall be provided with an emergency shutoff switch located within 100 feet (30480 mm) of, but not less than 20 feet (6096 mm) from, dispensers.

~~2.~~ 3. The owner of the LP-gas motor fuel-dispensing facility or the owner's designee shall provide for the safe operation of the system and the training of users.

4. The dispenser and hose-end valve shall release not more than ⅛ fluid ounce (4 cc) of liquid to the atmosphere upon breaking the connection with the fill valve on the vehicle.

2307.4 continues

2307.4 continued

5. Fire extinguishers shall be provided in accordance with Section 2305.4.

6. Warning signs shall be provided in accordance with Section 2305.6.

7. The area around the dispenser shall be maintained in accordance with Section 2305.7.

2307.8 Overfilling. LP-gas containers shall not be filled with LP-gas in excess of the volume determined using the fixed maximum liquid level gauge installed on the container, the volume determined by the overfilling prevention device installed on the container ~~outage installed by the manufacturer~~ or the weight determined by the required percentage of the water capacity marked on the container ~~stamped on the tank~~.

CHANGE SIGNIFICANCE: A number of revisions have been made to improve correlation of the requirements in the IFC with NFPA 58, *Liquefied Petroleum Gas Code*. Section 2307.4 has been reformatted into a user-friendly list of separate requirements. A revision included in this reformat changes the separation distance between LP-gas fuel-dispensing operations and the public way from 25 to 10 feet; that brings the IFC into agreement with NFPA 58 distance requirements.

The reference to Section 2306.7 (which addresses dispenser installations for gasoline and diesel fuels and did not make sense when applied to LP-gas installations) has been removed and replaced with a new Section 2307.5 containing only the requirements found in Section 2306.7, which are applicable to LP-gas operations.

New Section 2307.6.1 contains requirements for valves in the LP-gas system. Item 1 requires a device that will stop the flow of LP-gas when the dispenser is displaced. This is similar to a shear valve required for gasoline dispensers. Item 2 requires both an excess flow valve and a manual shutoff valve when the dispenser is remotely located from the storage tank. Item 3 requires either an excess flow valve or an emergency shutoff valve where the dispensing hose connects to the liquid piping. Item 4 requires a listed automatic-closing type hose nozzle valve on the dispensing hose when the dispenser is located on a fueling island.

New Section 2707.6.3 requires a breakaway device to be located somewhere in the dispensing hose or at the hose connections at either end. The location of the device is not specified, but performance criteria stipulate that the dispenser must not be dislodged during the function of the breakaway device.

Probably the most significant of all the revisions occurs in Section 2307.7. While the 2012 IFC prohibited self-service LP-gas dispensing operated by the public, the 2015 IFC now allows public refueling from LP-gas dispensers. This revision has occurred in response to the new technologies available to safeguard refueling operations.

LP-gas refueling technology now provides the following features:

• Liquid transfer will not occur unless the hose-end valve is completely connected and securely in place on the fill valve of the vehicle. Section 2307.6.1, Item 4 requires an automatic-closing hose nozzle valve.

• Specific LP-gas connections are required for motor vehicle fueling operations.

- LP-gas connections are designed to release no more than 4 cm^3 of liquid to the atmosphere. Section 2307.7, Item 4 allows a maximum release of 4cc (0.8 teaspoon) of liquid.

- The LP-gas refueling system is a closed system, which means that there is no opportunity for air, water or any other contaminant to enter the system or for gas to escape from the system.

- Individuals must be trained in order to use the dispensing equipment. This requirement is facilitated by the use of key, code and card-lock dispensing systems. Only trained individuals are issued the necessary security devices to enable the refueling of the vehicle. The training of the end user is critical. The fire inspector should ask for the training log at these facilities.

- Many new LP-gas fueled vehicles are equipped with an overfill prevention device. Section 2307.8 allows for this as a method of regulating the maximum fill amount in dispensing operations.

3103.9.1

Structural Design of Multistory Tents and Membrane Structures

CHANGE TYPE: Addition

CHANGE SUMMARY: Temporary multistory tents and membrane structures are now required to comply with the structural requirements in the IBC.

2015 CODE: **3103.9.1 Tents and Membrane Structures Exceeding One Story.** Tents and membrane structures exceeding one story shall be designed and constructed to comply with Chapter 16 of the *International Building Code.*

CHANGE SIGNIFICANCE: Even though the provision of "temporary" limits the usability of these structures to less than 180 days, improper structural design can still lead to problems such as risks to life and safety. Many of these temporary tents and temporary membrane structures have multiple floors and are over 30 feet in height. As the height increases, the

Photo courtesy of HTS-USA

Two-story tent at the America's Cup Race

Photo courtesy of Losberger-USA

Interior of two-story tent

potential for collapse increases. Often these multistory structures are used at exhibits, expos, concerts, outdoor functions and fairs as part of a booth or display. The public is walking in, around and through the various levels of the structure.

The safety of the occupants is paramount, and since these structures are temporary, they do not go through a full building-code structural analysis and review. This new section includes a requirement for temporary multi-story tents and membrane structures to comply with the structural requirements of IBC Chapter 16.

This section provides the fire code official with a solid tool to use to verify proper structural design. However, most fire code officials do not routinely review structural integrity. This is most commonly performed by the building official and staff. But the requirement for a construction permit is in the IFC since this is a temporary structure. Therefore, enforcement of this requirement may take some coordination between the building official and the fire code official prior to the issuance of the IFC construction permit.

3105

Temporary Stage Canopies

CHANGE TYPE: Addition

CHANGE SUMMARY: Temporary stage canopies are now permitted and regulated under Chapter 31 and must have a structurally sound design.

2015 CODE:

SECTION 3101
GENERAL

3101.1 Scope. Tents, <u>temporary stage canopies</u> and membrane structures shall comply with this chapter. The provisions of Section 3103 are applicable only to temporary tents and membrane structures. The provisions of Section 3104 are applicable to temporary and permanent tents and membrane structures. Other temporary structures shall comply with the *International Building Code*.
(Sections not shown did not change.)

SECTION 3105
TEMPORARY STAGE CANOPIES

3105.1 General. Temporary stage canopies shall comply with Section 3104, Sections 3105.2 through 3105.8 and ANSI E1.21.

3105.2 Approval. Temporary stage canopies in excess of 400 square feet shall not be erected, operated or maintained for any purpose without first obtaining approval and a permit from the fire code official and the building official.

3105.3 Permits. Permits shall be required as set forth in Sections 105.6 and 105.7.

3105.4 Use Period. Temporary stage canopies shall not be erected for a period of more than 45 days.

3105.5 Required Documents. All of the following documents shall be submitted to the fire code official and the building official for review before a permit is approved:

1. Construction documents: Construction documents shall be prepared in accordance with the *International Building Code* by a registered design professional. Construction documents shall include:
   1.1 A summary sheet showing the building code used, design criteria, loads and support reactions.
   1.2 Detailed construction and installation drawings.
   1.3 Design calculations.
   1.4 Operating limits of the structure explicitly outlined by the design professional including environmental conditions and physical forces.

Temporary stage canopy at an outdoor setting

Photo courtesy of Anne vonWeller

Temporary stage canopy after structural collapse at the Indiana State Fair, August 13, 2011

      1.5 Effects of additive elements such as video walls, supported scenery, audio equipment, and vertical and horizontal coverings.

      1.6 Means for adequate stability including specific requirements for guying and cross-bracing, ground anchors or ballast for different ground conditions.

  2. Designation of responsible party: The owner of the temporary stage canopy shall designate in writing a person to have responsibility for the temporary stage canopy on the site. The designated person shall have sufficient knowledge of the construction documents, manufacturer's recommendations and operations plan to make judgments regarding the structure's safety and to coordinate with the fire code official.

  3. Operations plan: The operations plan shall reflect manufacturer's operational guidelines, procedures for environmental monitoring and actions to be taken under specified conditions consistent with the construction documents.

3105.6 Inspections. Inspections shall comply with Section 106 and Sections 3105.6.1 and 3105.6.2.

3105.6.1 Independent Inspector. The owner of a temporary stage canopy shall employ a qualified, independent approved agency or individual to inspect the installation of a temporary stage canopy.

3105.6.2 Inspection Report. The inspecting agency or individual shall furnish an inspection report to the fire code official. The inspection report shall indicate that the temporary stage canopy was inspected and was or was not installed in accordance with the approved construction documents. Discrepancies shall be brought to the immediate attention of the installer for correction. Where any discrepancy is not corrected, it shall be brought to the attention of the fire code official and the designated responsible party.

3105.7 Means of Egress. The means of egress for temporary stage canopies shall comply with Chapter 10.

3105.8 Location. Temporary stage canopies shall be located a distance from property lines and buildings to accommodate distances indicated in the construction drawings for guy wires, cross-bracing, ground anchors or ballast. Location shall not interfere with egress from a building or encroach on fire apparatus access roads.

SECTION 202
GENERAL DEFINITIONS

Temporary Stage Canopy. A temporary ground-supported membrane-covered frame structure used to cover stage areas and support equipment in the production of outdoor entertainment events.

3105 continues

3105 continued

CHAPTER 80
REFERENCED STANDARDS

ANSI

E1.21-2006 Entertainment Technology: Temporary Ground Supported Overhead Structures Used to Cover the Stage Areas and Support Equipment in the Production of Outdoor Entertainment Events . . . 3105.1

CHANGE SIGNIFICANCE: Provisions have been added to the IFC addressing temporary stage canopies. These structures are constructed either on or around a stage to provide shelter; to support lighting and speakers; to display banners and advertisements; and to conceal equipment, props and scenery from the audience view. The size of the temporary stage canopy increases as the size of the concert or performance grows. Some temporary stage canopies can be over 50 feet in height.

There have been several high-profile temporary stage canopy collapses in the past few years:

1. August 7, 2011, at the Brady District Block Party, Tulsa, Oklahoma
2. July 17, 2011, at the Cisco Ottawa Blues Festival, Ottawa, Canada
3. August 13, 2011, at the State Fairgrounds, Indianapolis, Indiana (seven fatalities and over 50 injured)
4. August 18, 2011, at the Pukkelpop Festival, Kiewit, Belgium
5. June 16, 2012, at the Radiohead concert, Toronto, Canada (one fatality).

All of these collapses resulted in significant property damage, and two collapses resulted in fatalities. The obvious concern is for the safety of the performers and audiences along with behind-the-scene workers—stagehands, lighting technicians, security personnel and every other individual in proximity to the stage.

Temporary stage canopies are very complex and specialized for each individual event. The nature of these structures must accommodate a wide variety of changing components such as audio and lighting equipment, video walls and scenery. The entertainment industry is continually evolving new ways to improve shows creating larger, more complex and more spectacular presentations.

Due to the unique design of temporary stage canopies, it is difficult for most fire inspectors to find adequate guidance in current code language to satisfactorily regulate these specialized structures.

The regulations apply when the area of the temporary stage canopy exceeds 400 square feet. One difference when compared to other temporary structures regulated in IFC Chapter 31 is that the time period considered as "temporary" in Section 3105.4 is a maximum of 45 days. For structures to be used longer than that period they would be treated as permanent and regulated by the IBC.

One of the fundamental issues is the structural design of the temporary stage canopy. For this reason, Section 3105.2 requires that both the fire code official and the building official approve this system. Even though the temporary stage canopy is not a permanent building, it must comply with the structural requirements in the IBC for permanent structures. Review of structural design is typically conducted by the building official, so this section intends to continue with that typical review process.

A new ANSI standard is referenced in Section 3105.1, which specifically addresses the structural design and stability of temporary stage canopies. ANSI Standard E1.21-2006, Entertainment Technology: Temporary Ground Supported Overhead Structures Used to Cover the Stage Areas and Support Equipment in the Production of Outdoor Entertainment Events was produced by the Entertainment Services and Technology Association (ESTA). ESTA recently merged with an international organization, Professional Lighting and Sound Association (PLASA). The standard can be downloaded at no cost at www.plasa.org.

ANSI E1.21 includes requirements that the temporary stage canopy be designed not only to support the weight of the equipment but also to withstand seismic loads, wind loads, rain and snow loading, and uplift. There is a necessity for a design professional to fully analyze and design the structure to comply with the structural load requirements. Section 3.4 of ANSI E1.21 requires engineering drawings and calculations for the design, allowable payload, and maximum wind speed during erection process and also during use of the structure.

IFC Section 3105.5 Item 3 and ANSI E1.21 both require an operations plan to address the varying equipment weight for different shows, the need to monitor changing weather conditions, and safety operations when raising or lowering the roof to install equipment. ANSI E1.21 requires that an individual be designated to be responsible for the erection, use and operation of the temporary stage canopy. The operations plan must identify the responsible person, designate safe evacuation routes and locations, and evaluate the potential for the show being cancelled when the weather creates an unsafe condition.

Section 3105.6.1 requires an independent and qualified inspector or inspection agency to inspect the construction of the temporary stage canopy. The Professional Light and Sound Association (PLASA) offers an Entertainment Technician Certification Program (ETCP). This certification demonstrates the competence of the certificate holder to inspect temporary stage canopies. For large, unusually complex canopies there is latitude for the fire code official to require inspection by a structural engineer familiar with these types of temporary structures. Although it is not specified in the code, IBC Chapter 17, Special Inspections and Tests could possibly be used as a guide for this particular inspection requirement. The use of a special inspector could be a condition of approval from the building official.

3203.2
Class I Commodities

CHANGE TYPE: Modification

CHANGE SUMMARY: A building containing Class I commodities stored on plastic pallets will now require a fire sprinkler system to be designed based on the NFPA 13 sprinkler criteria. The allowance to include any solid-deck polyethylene pallets as acceptable for Class I commodities has been deleted.

2015 CODE: 3203.2 Class I Commodities.   Class I commodities are essentially noncombustible products on wooden ~~or nonexpanded polyethylene solid deck~~ pallets, in ordinary corrugated cartons with or without single-thickness dividers, or in ordinary paper wrappings with or without pallets. Class I commodities are allowed to contain a limited amount of Group A plastics in accordance with Section 3203.7.4. Examples of Class I commodities include, but are not limited to, the following:

Alcoholic beverages not exceeding 20-percent alcohol

Appliances noncombustible, electrical

Cement in bags

Ceramics

Dairy products in nonwax-coated containers (excluding bottles)

Dry insecticides

Foods in noncombustible containers

Fresh fruits and vegetables in nonplastic trays or containers

Frozen foods

Glass

Glycol in metal cans

Gypsum board

Typical Class I commodity: noncombustible food product in metal cans in single-thickness cardboard box on a wooden pallet

Inert materials, bagged

Insulation, noncombustible

Noncombustible liquids in plastic containers having less than a 5-gallon (19 L) capacity

Noncombustible metal products

CHANGE SIGNIFICANCE: The classification of commodities is used to determine the level of many fire safety features. The 2012 IFC includes solid-deck polyethylene pallets as an acceptable pallet for classification as a Class I commodity. There have been several fires of Class I commodities stored on plastic pallets where the fire sprinkler system, designed for a Class I commodity, was not able to control the fire. This is documented in a report by Scott Stookey (former ICC Senior Technical Staff in Product Development) that points to the possible failure of the fire sprinkler system since the protection of noncombustible materials on nonexpanded polyethylene solid deck was based on Class I Commodity.

With regard to fire protection, there are basically two types of plastic pallets: plastic pallets that have been tested and listed, and plastic pallets that are not listed. The listing for plastic pallets essentially demonstrates that the fire load added by a plastic pallet is equivalent to the fire load added by a wooden pallet. See the Significant Change to Section 3206.4.1 for additional information on plastic pallets.

According to Section 5.6.2.1 of NFPA 13, the fire hazard of the pallet is considered and included when classifying commodities, but only for the following types of pallets:

- Wood pallets,
- Metal pallets and
- Pallets that are listed as equivalent to wood pallets.

However, IFC Section 102.7.1 states that when the provisions in the IFC conflict with the provisions in the referenced standard, the IFC language takes precedence. In this case, NFPA 13 considered only listed plastic pallets as part of the commodity classification as Class I, but the IFC allowed the use of any nonexpanded polyethylene solid-deck pallet to be part of the Class I commodity classification. Therefore, even though NFPA 13 considered only "listed plastic pallets" as part of the commodity classification as Class I, the IFC language overruled this provision and allowed plastic pallets that were not listed. This resulted in a fire sprinkler system design that was below the required discharge density to handle the fire load.

In NFPA 13, Standard for the Installation of Sprinkler Systems, plastic pallets are not considered equivalent to wood pallets unless listed as such. Section 3206.4.1.1of the 2015 IFC now specifies that plastic pallets will be considered as equivalent to wood pallets when listed as complying with the test standard. (See the Significant Change to IFC Section 3206.4.1.) The 2015 IFC is now consistent with the NFPA 13 requirements.

According to NFPA 13 Section 5.6.2.2, the use of unreinforced polypropylene or unreinforced high-density polyethylene plastic pallets will increase the commodity classification by one classification level.

3203.2 continues

3203.2 continued

Therefore, when a Class I commodity is stored on a unlisted plastic pallet, whether it is solid deck or not, it will be classified as a Class II commodity, and the fire sprinkler system will be designed for protection of a Class II commodity.

NFPA 13 goes on to say in Section 5.6.2.3 that "reinforced polypropylene" or "reinforced high-density polyethylene" plastic pallets will increase the commodity classification by two classification levels, except for Class IV commodities, which will be increased to a Group A cartoned, unexpanded plastic commodity. A reinforced pallet is, by definition, a plastic pallet that incorporates steel or fiberglass for increased structural support and stability.

A designer must consider the increase in commodity classification for unlisted plastic pallets and unlisted reinforced plastic pallets when designing the fire sprinkler system. The increased classification may also have an effect when Table 3206.2 is applied for the other fire safety features in high-piled storage areas.

CHANGE TYPE: Modification

CHANGE SUMMARY: NFPA 13 provisions are now referenced to address the use of plastic pallets in high-piled combustible storage. Plastic pallets can affect the classification of the commodity.

2015 CODE: **3206.4.1 Pallets.** Automatic sprinkler system requirements based upon the presence of pallets shall be in accordance with NFPA 13.

3206.4.1.1 Plastic Pallets. Plastic pallets listed and labeled in accordance with UL 2335 or FM 4996 shall be treated as wood pallets for determining required sprinkler protection.

3208.2.1 Plastic ~~Pallets and~~ Shelves. Storage on ~~plastic pallets or~~ plastic shelves shall be protected by approved specially engineered fire protection systems.

> **~~Exception:~~** ~~Plastic pallets listed and labeled in accordance with UL 2335 shall be treated as wood pallets for determining required sprinkler protection.~~

CHAPTER 80
REFERENCED STANDARDS

FM

ANSI/FM 4996-13 Standard for Classification of Pallets and Other Material Handling Products as Equivalent to Wood Pallets . . . 3206.4.1.1

CHANGE SIGNIFICANCE: The 2012 IFC contained the reference to listed plastic pallets in Section 3208.2.1. The location of this section implied that plastic pallets were only a concern when used in rack storage. However, the fire hazard associated with plastic pallets occurs whether they are used in rack storage or palletized storage. Therefore, the first main change in the 2015 edition is that the issue of plastic pallets is

3206.4.1 continues

3206.4.1
Plastic Pallets Used in High-piled Combustible Storage

Plastic pallet listed under UL 2335 to be considered equivalent to a wood pallet

Photo courtesy of Rehrig Pacific Company, Los Angeles, CA

3206.4.1 continued

relocated to Section 3206.4. This section applies to any place that pallets could be used in a storage configuration. See also the Significant Change to Section 3203.2, which discusses the removal of the reference to plastic pallets from Class I commodities.

More important though is the requirement that fire sprinkler design is based on the requirements contained in NFPA 13, Standard for the Installation of Sprinkler Systems. NFPA 13 has extensive requirements for pallet use and storage based on the type of pallet.

NFPA 13 Section 5.6.2.1 states that the fire hazard of the pallet is considered and included when commodities are classified, but only when the pallets are wood, metal or pallets that are listed as equivalent to wood pallets.

NFPA 13 Section 5.6.2.2 requires that the commodity classification is to be increased by one classification level when unreinforced polypropylene or unreinforced high-density polyethylene plastic pallets are used.

NFPA 13 Section 5.6.2.3 requires that the commodity classification is to be increased by one classification level when "reinforced polypropylene" or "reinforced high-density polyethylene" plastic pallets are used. One exception to this is when the commodity is already Class IV, then the classification is to be increased to a Group A cartoned, unexpanded plastic commodity. A reinforced pallet is, by definition, a plastic pallet that incorporates steel or fiberglass for increased structural support and stability.

NFPA 13 Section 5.6.2.3.1 states that pallets are to be assumed as reinforced plastic pallets unless a permanent marking or manufacturer's certification is provided. This is based on the fact that the reinforced pallets create a higher fire sprinkler demand, so the assumption to go the higher hazard is appropriate unless it can be proven otherwise.

NFPA 13 Section 5.6.2.7 states that when pallets are other than wood, metal, polypropylene plastic or polyethylene plastic that the commodity classification will be based on either specific testing or increased two commodity classifications.

IFC Section 3206.4.1.1 now references two test standards with regard to accepting plastic pallets. These two standards are

1. UL 2335-2010, Fire Tests of Storage Pallets—with Revisions through September 2012 and
2. FM 4996-2013, Standard for Classification of Pallets and Other Material Handling Products as Equivalent to Wood Pallets.

UL 2335 is referenced in the 2012 IFC. FM 4996 has been included in the 2015 IFC to allow an alternative. Either standard can be utilized for the testing of plastic pallets.

When polyethylene or polypropylene pallets are used as part of the method of storage, the fire hazard of the plastic pallet must be evaluated and included in the design of the fire sprinkler system and other fire protection features in Table 3206.2. Table 3206-1 shows how the use of plastic pallets will affect the commodity classification.

TABLE 3206-1   Adjustments to the Commodity Classification Based on the Use of Plastic Pallets

Type of Plastic Pallet	Listed under UL 2335, or Approved under FM 4996	Non listed or Non approved Plastic Pallet
Unreinforced polyethylene or polypropylene	Treat as equivalent to a wood pallet	Increase one commodity classification
Reinforced polyethylene or polypropylene	Treat as equivalent to a wood pallet	For Class I, II or III commodities, increase two commodity classifications For Class IV commodities, treat as a Group A cartoned, unexpanded plastic commodity
Plastic materials other than polyethylene or polypropylene	Treat as equivalent to a wood pallet	Commodity classification based on specific fire testing, or increase two commodity classifications

3206.9.3

Dead-end Aisles in High-piled Combustible Storage

CHANGE TYPE: Clarification

CHANGE SUMMARY: Specific limitations are now provided for dead-end aisles in high-piled combustible storage areas. These limitations are more restrictive than the common path of egress travel limitations due to hazards associated with high-piled combustible storage.

2015 CODE: 3206.9.3 Dead-End Aisles. ~~Dead-end aisles shall be in accordance with Chapter 10.~~ Dead-end aisles shall not exceed 20 feet (6096 mm) in length in Group M Occupancies. Dead-end aisles shall not exceed 50 feet (15240 mm) in length in all other occupancies.

Exception: Dead-end aisles are not limited where the length of the dead-end aisle is less than 2.5 times the least width of the dead-end aisle.

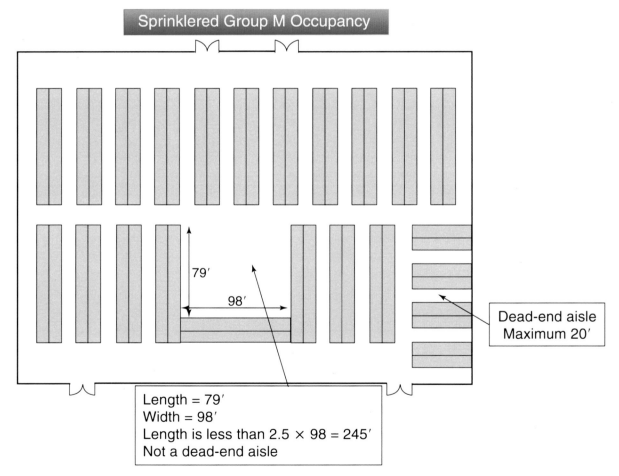

Sprinklered Group M Occupancy

79'

98'

Dead-end aisle Maximum 20'

Length = 79'
Width = 98'
Length is less than 2.5 × 98 = 245'
Not a dead-end aisle

Areas where the length is less than 2.5 times the width are not considered dead-end aisles.

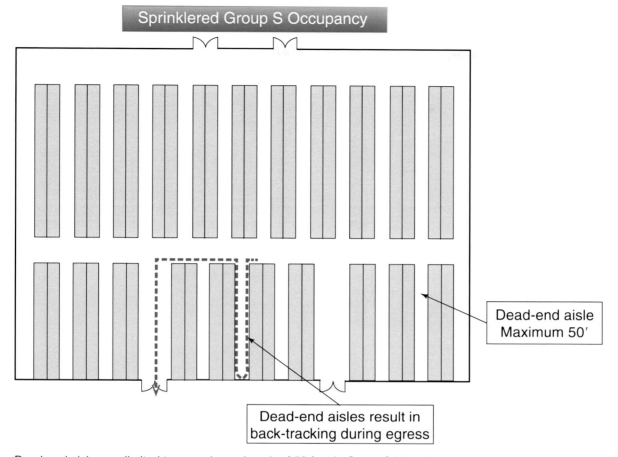

Dead-end aisles are limited to a maximum length of 50 feet in Group S high-piled storage areas.

CHANGE SIGNIFICANCE: The 2012 IFC refers to Chapter 10 for requirements for dead-end aisles in high-piled storage areas. But the only requirements in Chapter 10 are as follows:

1. Dead-ends in Section 1020.4, but this applies only to corridors.
2. Aisles in Group B and M occupancies in Section 1018, but there are no criteria for dead-ends.
3. Dead-ends in Section 1029.9.5 Exception 1, but this applies only to assembly occupancies.

Essentially, the anticipated requirements applicable to dead-end aisles in a warehouse or high-piled storage area did not exist. The only provision that could apply was Table 1006.2.1 for the common path of travel. This table allows for up to 100 feet of common path of egress travel, and this is excessive when considering the type of fire load found within high-piled combustible storage areas.

The limitation of 20 feet for dead-end aisles in Group M occupancies with high-piled storage is because the public is present. Group M occupancies with high-piled storage can have occupant loads over 1,000 persons. Based on the high potential for life loss, the limitation of 20 feet is justified.

In all other occupancies, which are most likely Group S, the dead-end aisles have a limit of 50 feet. This is similar to the limit of 50 feet in Group S in Section 1020.4 for dead-end corridors.

3206.9.3 continues

3206.9.3 continued   The foundational issue with regard to dead-end aisles is to answer the question: when an occupant is trying to exit a building in a fire situation and turns down a dead-end aisle, how far should an occupant need to travel before having to back-track? The longer the dead end is, the further the occupant is allowed to travel before turning back toward the exit.

Dead-end aisles only increase confusion and add to the evacuation time, both of which put the occupant in greater danger. The code does not prohibit dead-end aisles, but the 2015 IFC contains specific limitations to the length of the dead-end aisle.

The exception allows for areas that are not really aisles due to the width of the aisle compared to the length. When the length of the aisle is less than 2½ times the width of the aisle, then the area is not treated as a dead-end aisle. This is consistent with the concept in IFC Section 1018.4 with regard to dead-end corridors. Basically, the area is wide enough that it is not really a corridor, and therefore it is not a dead-end. Similarly, in high-piled storage areas, where the length does not exceed 2.5 times the width, then it is not really an aisle. It would simply be considered a sales area.

CHANGE TYPE: Addition

CHANGE SUMMARY: Safety requirements for the purging and cleaning of flammable gas piping systems have been added to the 2015 IFC.

2015 CODE: **3306.2 Cleaning with Flammable Gas.**   Flammable gases shall not be used to clean or remove debris from piping open to the atmosphere.

3306.2.1 Pipe Cleaning and Purging.   The cleaning and purging of flammable gas piping systems, including cleaning new or existing piping systems, purging piping systems into service, and purging piping systems out of service shall comply with NFPA 56.

Exceptions:

1. Compressed gas piping systems other than fuel gas piping systems where in accordance with Chapter 53.
2. Piping systems regulated by the *International Fuel Gas Code.*
3. Liquefied petroleum gas systems in accordance with Chapter 61.

**CHAPTER 80
REFERENCED STANDARDS**

NFPA
56-2012 Standard for Fire and Explosion Prevention during Cleaning and Purging of Flammable Gas Piping Systems . . . 3306.2.1

3306.2 continues

3306.2
Cleaning with Flammable Gas

Pipe cleaning and purging operation using flammable gas. Welding slag, pipe cuttings, threading burrs and debris should be cleaned from the piping prior to the piping being placed in service.

Photo courtesy of U.S. Chemical Safety Board

Pipe cleaning and purging operation using flammable gas. Pipe terminates inside the facility, surrounded by the process and equipment.

3306.2 continued

CHANGE SIGNIFICANCE: In the past few years, two separate explosions occurred that were attributed to workers using natural gas flowing at high velocities to clean and purge fuel-gas piping:

1. June 9, 2009, at the ConAgra Foods Slim Jim™ meat processing facility in Garner, North Carolina, four workers were killed and 67 others were injured in a natural gas explosion.
2. February 7, 2010, at the Kleen Energy power plant in Middletown, Connecticut, six workers were killed and at least 50 others were injured in a natural gas explosion.

Both of these incidents occurred during the commissioning of a fuel gas piping system. The new piping system was being cleaned to remove any dirt, debris, welding slag, etc. Flammable gas was used as the cleaning medium, and the gas and debris were expelled from the end of an open pipe terminating directly into the atmosphere. This is very similar to the flushing of the underground piping supplying a fire sprinkler system.

When new fuel gas piping is installed, debris such as rust, welding slag or other foreign material may have been introduced into the pipe. These materials can impact the fuel-burning equipment and clog orifices downstream if they remain after construction. Many equipment manufacturers require that the fuel gas supply piping is purged prior to connection of the equipment. Even when not required by the manufacturer, it is common practice to clean the piping prior to use.

The easy and convenient method of cleaning is to open the supply valve and have the flammable gas discharge at a velocity that is adequate to pick up and carry the debris to an open end of the piping system. The flammable gas and debris is discharged to the atmosphere and allowed to disperse. In the industry, this procedure is commonly referred to as a "gas blow."

The gas blow can be successful, as long as ignition does not occur. Fixed ignition sources can be controlled, but other methods of ignition are more difficult to control. An investigation by the U.S. Chemical Safety Board, conducted after the Kleen Energy incident, concluded other possible methods of ignition include metal debris carried by the flammable gas striking an object and creating a spark, or static electricity. For details on the incidents see the U.S. Chemical Safety Board report at: www.csb .gov/UserFiles/file/FINAL%20Urgent%20Recommendation.pdf

Both of these incidents involved the use of natural gas as the cleaning medium. Methane is the primary component of natural gas. Methane has a lower explosive limit (LEL) of 4.4 percent by volume and an upper explosive limit of 16.5 percent in air. Methane can readily form explosive mixtures that are easily ignited when mixed with air. An LEL of 4.4 percent creates a potentially hazardous situation with just a small quantity of gas.

Section 3306.2 specifically prohibits the use of flammable gas as the cleaning medium when the gas piping system will discharge to the atmosphere. There are other industry-accepted methods to clean fuel gas piping, such as the use of nitrogen or other inert gas, or the use of compressed air. These alternate methods do not pose a flammability hazard.

The NFPA 56, Standard for Fire and Explosion Prevention during Cleaning and Purging of Flammable Gas Piping Systems covers both new and existing gas piping systems at commercial facilities, industrial facilities and electric generating plants. The standard contains safety procedures and requirements for

- cleaning and purging of flammable gas piping systems,
- purging piping systems into service and
- purging piping systems out of service.

Two of the main requirements of NFPA 56 are that

1. written procedures must be developed for each cleaning or purging operation, and
2. flammable gas is not allowed to be used as the cleaning media.

3504.1.7, 3510

Hot Work on Flammable and Combustible Liquid Storage Tanks

CHANGE TYPE: Addition

CHANGE SUMMARY: Requirements for hot work on tanks containing flammable and combustible liquids is now included in the 2015 IFC.

2015 CODE: 3504.1.7 Precautions in Hot Work. Hot work shall not be performed on containers or equipment that contain or have contained flammable liquids, gases or solids until the containers and equipment have been thoroughly cleaned, inerted or purged; except that "hot tapping" shall be allowed on tanks and pipe lines when such work is to be conducted by approved personnel. <u>Hot work on flammable and combustible liquid storage tanks shall be conducted in accordance with Section 3510.</u>

SECTION 3510
HOT WORK ON FLAMMABLE AND COMBUSTIBLE LIQUID STORAGE TANKS

<u>**3510.1 General.** Hot work performed on the interior or exterior of tanks that hold or have held flammable or combustible liquids shall be in accordance with Section 3510.2 and Chapters 4, 5, 6, 7 and 10 of NFPA 326.</u>

<u>**3510.2 Prevention.** The following steps shall be taken to minimize hazards where hot work must be performed on a flammable or combustible liquid storage container:</u>

1. <u>Use alternative methods to avoid hot work where possible.</u>
2. <u>Analyze the hazards prior to performing hot work, identify the potential hazards and the methods of hazard control.</u>

Hot work on tanks containing flammable and combustible liquids is specifically regulated in the IFC.

3. Hot work shall conform to the requirements of the code or standard to which the container was originally fabricated.

4. Test the immediate and surrounding work area with a combustible gas detector and provide for a means of continuing monitoring while conducting the hot work.

5. Qualified employees and contractors performing hot work shall use an industry-approved hot work permit system to control the work.

6. Personnel shall be properly trained on hot work policies and procedures regarding equipment, safety, hazard controls and job-specific requirements.

7. On-site safety supervision shall be present when hot work is in progress to protect the personnel conducting the hot work and provide additional overview of site specific hazards.

SECTION 202
GENERAL DEFINITIONS

Combustible Gas Detector. An instrument that samples the local atmosphere and indicates the presence of ignitable vapors or gases within the flammable or explosive range expressed as a volume percent in air.

CHAPTER 80
REFERENCED STANDARDS

NFPA

326-2010 Standard for the Safeguarding of Tanks and Containers for Entry, Cleaning or Repair . . . 3510.1

CHANGE SIGNIFICANCE: Annually, there are numerous incidents resulting from hot-work operations. Many of those incidents involve loss of life, injury and significant property damage. These code revisions improve the general guidelines in the IFC for safely conducting hot work on tanks containing flammable or combustible liquids and include a new reference to NFPA 326, Standard for the Safeguarding of Tanks and Containers for Entry, Cleaning or Repair.

Section 3504.1.7 states that hot work on containers and equipment shall only be performed after the container, piping or equipment has been cleaned, purged or otherwise safeguarded. It then goes on to state that hot tapping shall be performed by approved personnel only. Hot tapping is extremely dangerous since the piping or tank still contains flammable or combustible liquids, and therefore the personnel performing this work must be aware of the hazards and be able to safeguard the operation.

Hot tapping is the process of cutting or welding on a tank or piping while the tank or piping is still in service and the tank or piping contains flammable or combustible liquids (IFC Section 202).

The new reference in Section 3504.1.7 sends the code user to new Section 3510 for requirements concerning hot work conducted on flammable and combustible liquid tanks. Section 3510.1 in turn references NFPA 326 for these hot-work operations. NFPA 326 does not cover hot tapping; other industry standards and safeguards must be used.

3504.1.7, 3510 continues

3504.1.7, 3510 continued

The reference to NFPA 326 includes only specific chapters, not the entire standard. The referenced chapters cover

- Chapter 4 – Basic Precautions
- Chapter 5 – Preparation for Safeguarding
- Chapter 6 – Testing Procedures
- Chapter 7 – Control or Removal of Vapors
- Chapter 10 – Cleaning Tanks and Containers.

IFC Section 105.6.24 also requires a hot-work operations permit. For repetitive or routine hot-work operations, IFC Section 3503.3 allows for the development of a Hot Work Program whereby the facility is issued a blanket operations permit and individual hot-work operations are inspected, supervised and permitted by a designated responsible person at the facility.

Hot Work Program is a permitted program, carried out by *approved* facilities-designated personnel, allowing them to oversee and issue permits for hot work conducted by their personnel or at their facility. The intent is to have trained, on-site, responsible personnel ensure who required hot work safety measures are taken to prevent fires and fire spread (IFC Section 202).

The U.S. Chemical Safety Board has published general guidelines applicable to most every situation where welding or cutting on flammable, combustible or toxic tanks occur. Those general safety guidelines are now listed in Section 3510.2.

PART 5

Hazardous Materials

Chapters 50 through 79

Chapters 50 through 67 contain requirements for the safe storage, use, handling and dispensing of hazardous materials. The Chemical Abstract Service, a part of the American Chemical Council, has identified over 35 million organic and inorganic substances and over 60 million sequences or mixtures. The *International Fire Code* (IFC) does not regulate all of these chemicals. Instead, the IFC focuses on chemicals or chemical mixtures that pose physical or health hazards. Physical hazard materials can detonate, deflagrate, burn or accelerate burning. Health hazard materials can incapacitate, injure or cause death after only a single exposure. Chapters 68 through 79 are reserved and serve as place holders for future code provisions when new hazardous materials are identified as requiring regulation by the IFC. ■

TABLE 5003.1.1(1)
Maximum Allowable Quantities of Hazardous Materials

5101.4, 5104
Plastic Aerosol Containers

5307
Carbon Dioxide (CO$_2$) Systems Used in Beverage Dispensing Applications

5704.2.9.7.3
Flame Arresters on Protected Above-ground Storage Tanks

5808
Hydrogen Fuel Gas Rooms

CHANGE TYPE: Modification

CHANGE SUMMARY: Table 5003.1.1(1) contains several revisions affecting consumer fireworks, combustible fibers, unstable reactive materials, alcohol-based hand rubs and gas rooms.

2015 CODE: See below.

CHANGE SIGNIFICANCE: The 2015 IFC contains a number of editorial and format revisions to IFC Table 5003.1.1(1). These revisions are mirrored in IBC Table 307.1.1(1), so that the two tables are identical. Additional revisions occur in the specific criteria in Table 5003.1.1(1) in the following areas:

- Combustible Fibers
- Consumer Fireworks
- Unstable Reactive Materials
- Footnote c
- Footnote e
- Footnote p

There are several editorial revisions to Table 5003.1.1(1) that provide for consistency and uniformity in application of the provisions whether in the IBC or IFC. These revisions are as follows:

- All of the material classifications have been alphabetized.
- "Consumer Fireworks" has been relocated.

Table 5003.1.1(1) continues

Table 5003.1.1(1)

Maximum Allowable Quantities of Hazardous Materials

Consumer fireworks in a retail store

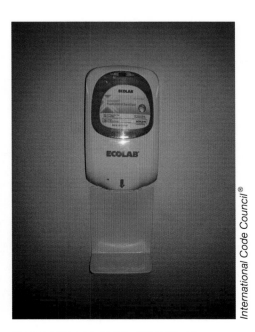

Flammable and combustible liquid quantities in alcohol-based hand-rub dispensers are not included when evaluating the MAQ for dispensers mounted and in use.

Table 5003.1.1(1) continued

TABLE 5003.1.1(1) Maximum Allowable Quantity per Control Area of Hazardous Materials Posing a Physical Hazard [a, j, m, n, p]

MATERIAL	CLASS	GROUP WHEN THE MAXIMUM ALLOWABLE QUANTITY IS EXCEEDED	STORAGE[b] Solid pounds (cubic feet)	STORAGE[b] Liquid gallons (pounds)	STORAGE[b] Gas (cubic feet at NTP)	USE-CLOSED SYSTEMS[b] Solid pounds (cubic feet)	USE-CLOSED SYSTEMS[b] Liquid gallons (pounds)	USE-CLOSED SYSTEMS[b] Gas (cubic feet at NTP)	USE-OPEN SYSTEMS[b] Solid pounds (cubic feet)	USE-OPEN SYSTEMS[b] Liquid gallons (pounds)
Combustible dust	NA	H-2	Note q	NA	NA	Note q	NA	NA	Note q	NA
Combustible fiber[q]	Loose	H-3	(100)	NA	NA	(100)	NA	NA	(20)	NA
	Baled[o]		(1,000)			(1,000)			(200)	
Combustible liquid[c, i]	II	H-2 or H-3	NA	120[d, e]	NA	NA	120[d]	NA	NA	30[d]
	IIIA	H-2 or H-3	NA	330[d, e]	NA	NA	330[d]	NA	NA	80[d]
	IIIB	NA	NA	13,200[e, f]	NA	NA	13,200[f]	NA	NA	3,300[f]
Consumer fireworks	1.4G	H-3	125[d, e, l]	NA	NA	NA	NA	NA	NA	NA
Cryogenic Flammable	NA	H-2	NA	45[d]	NA	NA	45[d]	NA	NA	10[d]
Cryogenic Inert	NA	NA	NA	NA	NL	NA	NA	NL	NA	NA
Cryogenic Oxidizing	NA	H-3	NA	45[d]	NA	NA	45[d]	NA	NA	10[d]
Explosives	Division 1.1	H-1	1[e, g]	(1)[e, g]	NA	0.25[g]	(0.25)[g]	NA	0.25[g]	(0.25)[g]
	Division 1.2	H-1	1[e, g]	(1)[e, g]	NA	0.25[g]	(0.25)[g]	NA	0.25[g]	(0.25)[g]
	Division 1.3	H-1 or H-2	5[e, g]	(5)[e, g]	NA	1[g]	(1)[g]	NA	1[g]	(1)[g]
	Division 1.4	H-3	50[e, g]	(50)[e, g]	NA	50[g]	(50)[g]	NA	NA	NA
	Division 1.4G	H-3	125[d, e, l]	NA	NA	NA	NA	NA	NA	NA
	Division 1.5	H-1	1[e, g]	(1)[e, g]	NA	0.25[g]	(0.25)[g]	NA	0.25[g]	(0.25)[g]
	Division 1.6	H-1	1[e, g]	NA	NA	NA	NA	NA	NA	NA
Flammable gas	Gaseous	H-2	NA	NA	1,000[d, e]	NA	NA	1,000[d, e]	NA	NA
	Liquefied		NA	(150)[d, e]	NA	NA	(150)[d, e]	NA	NA	NA
Flammable liquid[c]	1A	H-2 or H-3	NA	30[d, e]	NA	NA	30[d]	NA	NA	10[d]
	1B and IC		NA	120[d, e]	NA	NA	120[d]	NA	NA	30[d]
Flammable liquid combination (1A, 1B, 1C)	NA	H-2 or H-3	NA	120[d, e, h]	NA	NA	120[d, h]	NA	NA	30[d, h]
Flammable solid	NA	H-3	125[d, e]	NA	NA	125[d]	NA	NA	25[d]	NA
Inert gas	Gaseous	NA	NA	NA	NL	NA	NA	NL	NA	NA
	Liquefied	NA	NA	NA	NL	NA	NA	NL	NA	NA
Cryogenic inert	NA	NA	NA	NA	NL	NA	NA	NL	NA	NA
Organic peroxide	UD	H-1	1[e, g]	(1)[e, g]	NA	0.25[g]	(0.25)[g]	NA	0.25[g]	(0.25)[g]
	I	H-2	5[d, e]	(5)[d, e]	NA	1[d]	(1)[d]	NA	1[d]	(1)[d]
	II	H-3	50[d, e]	(50)[d, e]	NA	50[d]	(50)[d]	NA	10[d]	(10)[d]
	III	H-3	125[d, e]	(125)[d, e]	NA	125[d]	(125)[d]	NA	25[d]	(25)[d]
	IV	NA	NL	NL	NA	NL	NL	NA	NL	NL
	V	NA	NL	NL	NA	NL	NL	NA	NL	NL

Material	Class	Group							
Oxidizer	4	H-1	1^{e,g}	NA	0.25^g	(0.25)^g	NA	0.25^g	(0.25)^g
	3^k	H-2 or H-3	10^{d,e}	NA	2^d	(2)^d	NA	2^d	(2)^d
	2	H-3	250^{d,e}	NA	250^d	(250)^d	NA	50^d	(50)^d
	1	NA	4,000^{e,f}	NA	4,000^f	(4,000)^f	NA	1,000^f	(1,000)^f
Oxidizing gas	Gaseous	H-3	NA	1,500^{d,e}	NA	NA	1,500^{d,e}	NA	NA
	Liquefied		150^{d,e}	NA	NA	(150)^{d,e}	NA	NA	NA
Pyrophoric	NA	H-2	4^{e,g}	50^{e,g}	1^g	(1)^g	10^{e,g}	0	0
Unstable (reactive)	4	H-1	1^{e,g}	10^g	0.25^g	(0.25)^g	2^{e,g}	0.25^g	(0.25)^g
	3	H-1 or H-2	5^{d,e}	50^{d,e}	1^d	(1)^d	10^{d,e}	1^d	(1)^d
	2	H-3	50^{d,e}	~~250~~ 750^{d,e}	50^d	(50)^d	~~250~~ 750^{d,e}	10^d	(10)^d
	1	NA	NL	NL	NL	NL	NL	NL	NL
Water reactive	3	H-2	5^{d,e}	NA	5^d	(5)^d	NA	1^d	(1)^d
	2	H-3	50^{d,e}	NA	50^d	(50)^d	NA	10^d	(10)^d
	1	NA	NL	NA	NL	NL	NA	NL	NL

For SI: 1 cubic foot = 0.02832 m³, 1 pound = 0.454 kg, 1 gallon = 3.785 L.

NA = Not Applicable; NL = Not Limited; UD = Unclassified Detonable.

a. For use of control areas, see Section 5003.8.3.

b. The aggregate quantity in use and storage shall not exceed the quantity listed for storage.

c. The quantities of alcoholic beverages in retail and wholesale sales occupancies shall not be limited providing the liquids are packaged in individual containers not exceeding 1.3 gallons. In retail and wholesale sales occupancies, the quantities of medicines, foodstuffs, or consumer or industrial products, and cosmetics containing not more than 50 percent by volume of water-miscible liquids with the remainder of the solutions not being flammable shall not be limited, provided that such materials are packaged in individual containers not exceeding 1.3 gallons.

d. Maximum allowable quantities shall be increased 100 percent in buildings equipped throughout with an approved automatic sprinkler system in accordance with Section 903.3.1.1. Where Note e also applies, the increase for both notes shall be applied accumulatively.

e. Maximum allowable quantities shall be increased 100 percent when stored in approved storage cabinets, day boxes, gas cabinets, gas rooms, exhausted enclosures or listed safety cans in accordance with Section 5003.9.10. Where Note d also applies, the increase for both notes shall be applied accumulatively.

f. Quantities shall not be limited in a building equipped throughout with an approved automatic sprinkler system.

g. Allowed only in buildings equipped throughout with an approved automatic sprinkler system.

h. Containing not more than the maximum allowable quantity per control area of Class IA, Class IB or Class IC flammable liquids.

i. The maximum allowable quantity shall not apply to fuel oil storage complying with Section 603.3.2.

j. Quantities in parentheses indicate quantity units in parentheses at the head of each column.

k. A maximum quantity of 200 pounds of solid or 20 gallons of liquid Class 3 oxidizers is allowed when such materials are necessary for maintenance purposes, operation or sanitation of equipment when the storage containers and the manner of storage are approved.

l. Net weight of the pyrotechnic composition of the fireworks. Where the net weight of the pyrotechnic composition of the fireworks is not known, 25 percent of the gross weight of the fireworks including packaging shall be used.

m. For gallons of liquids, divide the amount in pounds by 10 in accordance with Section 5003.1.2.

n. For storage and display quantities in Group M and storage quantities in Group S occupancies complying with Section 5003.11, see Table 5003.11.1.

o. Densely-packed baled cotton that complies with the packing requirements of ISO 8115 shall not be included in this material class.

p. The following shall not be included in determining the maximum allowable quantities:
1. Liquid or gaseous fuel in fuel tanks on vehicles.
2. Liquid or gaseous fuel in fuel tanks on motorized equipment operated in accordance with this code.
3. Gaseous fuels in piping systems and fixed appliances regulated by the International Fuel Gas Code.
4. Liquid fuels in piping systems and fixed appliances regulated by the International Mechanical Code.
5. Alcohol-based hand rubs classified as Class I or II liquids in dispensers that are installed in accordance with Sections 5705.5 and 5705.5.1. The location of the alcohol-based hand-rub (ABHR) dispensers shall be provided in the construction documents.

q. Where manufactured, generated or used in such a manner that the concentration and conditions create a fire or explosion hazard based on information prepared in accordance with Section 104.7.2.

Table 5003.1.1(1) continues

Table 5003.1.1(1) continued

- 'Cryogenic inert' materials have been separated from inert materials and relocated to the other cryogenic listings.
- "Not Applicable" has been replaced with "NA."
- "Not Limited" has been replaced with "NL."

Combustible Fibers: The main hazard of combustible fibers is the ignitability of the product with rapid flame spread over exposed material surfaces. IFC Chapter 37 deals specifically with the hazards of combustible fibers, which present much the same hazard as combustible dust. The main characteristics of the fire hazard are the small mass as compared to the surface area of each particle and the ability for the product to become air-borne.

In the 2012 IFC, Footnote q was added to apply to combustible dust and basically referred the user to an evaluation of each instance where combustible dusts may be created to determine if classification as a Group H occupancy would be appropriate. Combustible fibers will now be treated in much the same manner. Footnote q is now applicable to combustible fibers as well as combustible dust.

In the case of combustible fibers, the footnote applies for the entire row. So, if the technical report concludes that combustible fibers are not manufactured, generated or used in such a manner to create a fire or explosion hazard, then the MAQs listed do not apply. However, if the report indicates that a fire or explosion hazard could exist, then the limitations for storage, use-open or use-closed would apply.

Consumer Fireworks: In the 2012 IFC the maximum allowable quantity per control area (MAQ) for consumer fireworks without benefit of an automatic sprinkler system was 125 pounds net. The MAQ could be doubled when an automatic sprinkler system was installed. This was consistent with the requirements in NFPA 1124, *Code for the Manufacture, Transportation, Storage, and Retail Sales of Fireworks and Pyrotechnic Articles.*

The question has been raised about the appropriateness of the fire-sprinkler system increase without fire-test data to justify such an increase. The decision has been made to eliminate the application of the footnote until fire testing that justifies the increase is completed. As a result, the MAQ for retail establishments is limited to 125 pounds net.

Unstable Reactive Materials: The MAQ for gaseous Class 3 unstable reactive materials has changed from 250 cubic feet to 750 cubic feet. This revision correlates the IFC with quantities in NFPA 55, *Compressed Gases and Cryogenic Fluids Code.* NFPA 55 is already referenced by the IFC, so this revision creates consistency in the application of both documents.

Footnote C: The term "industrial products" can apply to many different products and has led to confusion in the application of this footnote. For example, a wholesaler of car and truck batteries containing sulfuric acid (e.g., exceeding the MAQs for toxic and corrosive liquids) would be exempt based on Footnote c in the 2012 IFC. Footnote c was originally intended to exempt materials in small containers with small amounts of water-miscible hazardous materials in supermarkets, pharmacies and retail occupancies. This would include products such as household bleach, make up, and face toners. The reference to many other possible industrial products has been eliminated.

Footnote E: Gas rooms have been added to Footnote e as an allowed storage method for increasing the MAQ. Section 5003.8.4 has also been revised to require that when gas rooms are provided to increase the MAQ, the gas rooms must be

1. protected by a fire sprinkler system,
2. provided with a ventilation system designed to operate at a negative pressure as compared to the surrounding portions of the building, and
3. separated from the remainder of the building as required by the IBC.

When the fire code official is considering the required separation for Item 3 above, he or she must remember that if the gas room is being used in accordance with this footnote, then the MAQ is not exceeded. Since the MAQ is not exceeded, this gas room would not be classified as a Group H occupancy. Hence, the determination of the fire-resistance rating of the separation will be based on one of the following:

1. If the gas room is one control area and there are other control areas in the same story, then Table 5003.8.3.2 will provide the required separation between control areas.
2. If the gas room has an occupancy classification different from the remainder of the building, then IBC Table 508.4 will provide the required fire-resistance rating.

Essentially, gas rooms provide an equivalent level of protection as a gas cabinet or exhausted enclosure. Now any of these methods of storage will allow a 100 percent increase in the MAQ.

Footnote P: The revision to Footnote p allows alcohol-based hand-rub dispensers to be excluded when the fire code official is determining the MAQ for flammable or combustible liquids. The footnote indicates that the dispensers must comply with Sections 5705.5 and 5705.5.1.

Alcohol-based hand-rub dispensers provide a major service by limiting the spread of germs and improving infection control. As a result, alcohol-based hand-rub dispensers can be found in all types of occupancies.

Item 5 in Footnote p states that the quantities "in dispensers" are not included when determining the MAQ. However, somewhere in the facility is a large supply of refill canisters for the dispensers. The supply of refills is not included in this exemption and would therefore be included when the MAQ is evaluated. Additionally, the requirements in Sections 5705.5 and 5705.5.1 do not address storage of the refills. Storage is regulated by Section 5704.3.4.2, which provides specific quantity limits based on the occupancy classification, and Section 5704.3.4.4 regarding storage of flammable and combustible liquids below the MAQ.

5101.4, 5104

Plastic Aerosol Containers

Plastic aerosol container of shaving foam
(Photo courtesy of Balea)

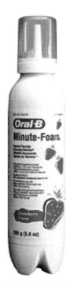

Oral B Minute-Foam Topical Flouride Foaming Solution in plastic aerosol container *(Photo courtesy of Proctor & Gamble Company, Cincinnati, OH)*

CHANGE TYPE: Modification

CHANGE SUMMARY: Aerosol products are now allowed in plastic containers up to 33.8 fluid ounces, or 1 liter, in size. Specific product criteria must be met if the plastic containers exceed 4 fluid ounces.

2015 CODE: <u>**5101.4 Containers.** Metal aerosol containers shall be limited to a maximum size of 33.8 fluid ounces (1000 ml). Plastic aerosol containers shall be limited to a maximum 4 fluid ounces (118 ml) except as provided in Section 5104.1.1. Glass aerosol containers shall be limited to a maximum 4 fluid ounces (118 ml).</u>

5104.1 General. The inside storage of Level 2 and 3 aerosol products shall comply with Sections 5104.2 through 5104.7and NFPA 30B. Level 1 aerosol products <u>and those aerosol products covered by Section 5104.1.1</u> shall be considered equivalent to a Class III commodity and shall comply with the requirements for palletized or rack storage in NFPA 13.

<u>**5104.1.1 Plastic Containers.** Aerosol products in plastic containers larger than 4 fluid ounces (118 ml), but not to exceed 33.8 fluid ounces (1000 ml) shall be allowed only where in accordance with this section. The commodity classification shall be Class III commodities as defined in NFPA 13 where any of the following conditions are met:</u>

<u>1. Base product has no fire point when tested in accordance with ASTM D 92, and nonflammable propellant.</u>
<u>2. Base product has no sustained combustion as tested in accordance with Appendix H, "Method of Testing for Sustained Combustibility in DOTn 49 CFR Part 173."</u>
<u>3. Base product contains up to 20 percent by volume (15.8 percent by weight) of ethanol and/or isopropyl alcohol in an aqueous mix, and nonflammable propellant.</u>
<u>4. Base product contains 4 percent by weight or less of an emulsified flammable liquefied gas propellant within an aqueous base. The propellant shall remain emulsified for the life of the product. Where such propellant is not permanently emulsified, the propellant shall be nonflammable.</u>

SECTION 202
GENERAL DEFINITIONS

Aerosol Container: A metal can, or a glass or plastic bottle designed to dispense an aerosol. ~~Metal cans shall be limited to a maximum size of 33.8 fluid ounces (1000 ml). Glass or plastic bottles shall be limited to a maximum size of 4 fluid ounces (118 ml).~~

CHANGE SIGNIFICANCE: These revisions affect aerosol products in plastic containers. The definition of "aerosol container" has been revised to delete specific size criteria. These criteria are now located in the new Section 5101.4.

New provisions in Section 5104.1.1 for aerosol products in plastic containers are based on new research and testing, which concludes that specific combinations of propellant and product can result in a low hazard product that does not present a significant fire hazard. If the aerosol product meets one of four criteria, the aerosol container can be as large as 1 liter, and the fire sprinkler design can be based on a Class III commodity.

To meet the requirements in Section 5104.1.1 item 1, the propellant must be nonflammable, and the product expelled must not have a fire point. The fire point would be tested to ASTM D92, Test Method for Flash and Fire Points by Cleveland Open Cup. A liquid can have a flash point, but not a fire point. An aerosol product that contains a nonflammable propellant and a liquid content that does not support combustion would have a Chemical Heat of Combustion of 0 kJ/g and be classified as a Level 1 aerosol product. Section 5104.1 already specifies that Level 1 aerosols be protected as a Class III commodity.

"Flash point" is the minimum temperature at which a liquid will give off sufficient vapors to form an ignitable mixture with air near the surface or in the container, but will not sustain combustion (IFC Section 202).

"Fire point" is the lowest temperature at which a liquid will ignite and achieve sustained burning when exposed to a test flame in accordance with ASTM D 92 (IFC Section 202).

Section 5104.1.1 item 2 provides for an aerosol container with a nonflammable propellant and a product that does not sustain combustion. The terminology is different compared to item 1 since the test standard in CFR 49 Appendix H, Method of Testing for Sustained Combustibility makes a determination of liquids which either "sustain combustion" or "do not sustain combustion." Essentially, Items 1 and 2 provide the same result; only a different test standard is utilized in each case.

Item 3 is based on specific fire testing conducted by FM Global. The alcohols used in the testing were ethanol and isopropyl alcohol. Therefore, item 3 is limited to only mixtures of ethanol or isopropyl alcohol. The mixture of 20 percent alcohol/80 percent water in a plastic bottle contained in cardboard cartons was tested in a full-scale array. Based on the test results, FM Global recommends protecting non-pressurized plastic bottles containing 20 percent$_{vol}$ alcohol/80 percent$_{vol}$ water mixtures with the same protection recommended for liquids that do not burn in plastic containers, i.e., Class I commodities.

Item 4 addresses aerosol containers with an emulsion of flammable liquefied gas propellant and an aqueous base. An emulsion in an aerosol product would be a mixture of two or more liquids in which one is present as droplets of microscopic or ultramicroscopic size distributed throughout the other. Emulsions are formed from the component liquids either spontaneously or, more often, by mechanical means, such as agitation, provided that the liquids that are mixed have no (or a very limited) mutual solubility. A good example is shaving cream.

The research used small-, intermediate- and full-scale fire testing to evaluate the fire hazard created by this type of product. During the fire tests, the liquid product was released and did not contribute to the fire, but the flammable liquefied gas did create brief flare-ups when released and created small fireballs throughout the test. The results of the intermediate-scale testing, the full-scale testing and the small-scale testing indicate that an aerosol product in a plastic container filled with a liquid mixture that does not support combustion and no more than 4 percent by weight flammable liquefied gas in a stable emulsion with the liquid mixture can be protected using criteria recommended for a Class III commodity.

5307

Carbon Dioxide (CO₂) Systems Used in Beverage Dispensing Applications

CHANGE TYPE: Addition

CHANGE SUMMARY: Large refrigerated carbon dioxide systems create a life safety hazard. Regulation of these systems is now included in the 2015 IFC.

2015 CODE:

SECTION 5307
CARBON DIOXIDE (CO₂) SYSTEMS USED IN BEVERAGE DISPENSING APPLICATIONS

5307.1 General. Carbon dioxide systems with more than 100 pounds (45.4 kg) of carbon dioxide used in beverage dispensing applications shall comply with Sections 5307.2 through 5307.5.2.

5307.2 Permits. Permits shall be required as set forth in Section 105.6.

5307.3 Equipment. The storage, use and handling of liquid carbon dioxide shall be in accordance with Chapter 53 and the applicable requirements of NFPA 55, Chapter 13. Insulated liquid carbon dioxide systems shall have pressure relief devices vented in accordance with NFPA 55.

5307.4 Protection From Damage. Carbon dioxide systems shall be installed, so the storage tanks, cylinders, piping and fittings are protected from damage by occupants or equipment during normal facility operations.

5307.5 Required protection. Where carbon dioxide storage tanks, cylinders, piping and equipment are located indoors, rooms or areas containing carbon dioxide storage tanks, cylinders, piping and fittings and

Exterior fitting to fill the bulk CO₂ cylinder inside the restaurant

Bulk CO₂ cylinder used for carbonated beverage dispensing system

other areas where a leak of carbon dioxide can collect shall be provided with either ventilation in accordance with Section 5307.5.1 or an emergency alarm system in accordance with Section 5307.5.2.

5307.5.1 Ventilation.   Mechanical ventilation shall be in accordance with the *International Mechanical Code* and shall comply with all of the following:

1. Mechanical ventilation in the room or area shall be at a rate of not less than 1 cubic foot per minute per square foot [0.00508 m^3/(s • m^2)].

2. Exhaust shall be taken from a point within 12 inches (305 mm) of the floor.

3. The ventilation system shall be designed to operate at a negative pressure in relation to the surrounding area.

5307.5.2 Emergency Alarm System.   An emergency alarm system shall comply with all of the following:

1. Continuous gas detection shall be provided to monitor areas where carbon dioxide can accumulate.

2. The threshold for activation of an alarm shall not exceed 5,000 parts per million (9,000 mg/m^3).

3. Activation of the emergency alarm system shall initiate a local alarm within the room or area in which the system is installed.

105.6.4 Carbon Dioxide Systems Used in Beverage Dispensing Applications.   An operational permit is required for carbon dioxide systems used in beverage dispensing applications with more than 100 pounds of carbon dioxide.

CHANGE SIGNIFICANCE: This new section addresses the use of liquid carbon dioxide (CO$_2$) in conjunction with carbonators to produce carbonated beverages. Many larger restaurants are converting from cylinders containing gaseous CO$_2$ and taking advantage of liquefied CO$_2$ because the reduced container size and reduced number of cylinders free up valuable floor space.

However, there is a hazard created by these systems that has resulted in a number of fatalities from CO$_2$ exposure in restaurants. The CO$_2$ can leak from large storage tanks, and it will displace oxygen in the room where the CO$_2$ container is located. CO$_2$ is an odorless and colorless gas with a vapor density of 1.53. The vapor density is 50 percent heavier than air, so the CO$_2$ will fill a room from the floor up and displace the oxygen in the room. The displacement of oxygen is more severe in smaller rooms or enclosures.

Carbon dioxide is stored as a compressed gas or a liquefied compressed gas. CO$_2$ is not classified as a cryogenic fluid because its boiling point is −109°F.

A trigger of 100 pounds has been set for application of these provisions. Smaller systems do not pose as great a risk of an exposed person's asphyxiation as compared with the larger quantities of CO$_2$. Section 5307.2

5307 continues

5307 continued provides a reference to operational permit requirements. The new operational permit in Section 105.6.4 also has 100 pounds as the threshold for an operational permit.

Components in a compressed gas system are already required to comply with Chapter 53, which will cover items such as the pressure vessel and piping requirements. There is also a specific reference to Chapter 13 of NFPA 55, *Compressed Gases and Cryogenic Fluids Code*. Chapter 13 addresses insulated liquid CO_2 systems in indoor and outdoor locations consisting of containers with a capacity of 1,000 pounds or less. The requirements address

- pressure relief devices,
- pressure and level indicators,
- piping systems including materials of construction and
- operating instructions.

A CO_2 release can accumulate and displace oxygen, creating an asphyxiation hazard. To prevent a potential mishap, the CO_2 storage tanks, cylinders, piping and fittings must be protected from physical damage. However, should a leak occur, Section 5307.5 requires one of two safeguards:

1. A mechanical ventilation system that exhausts to the exterior at a minimum rate of 1 cfm/ft². The ventilation intakes must be located within 12 inches of the floor since CO_2 is heavier than air.

2. An emergency alarm system consisting of gas detectors and a local alarm within the room where the CO_2 is stored. The trigger level of 5,000 ppm is the 8-hour time-weighted average exposure limit for CO_2 established by the United States Department of Labor, Occupational Safety and Health Administration (OSHA) for CO_2.

5704.2.9.7.3

Flame Arresters on Protected Above-ground Storage Tanks

CHANGE TYPE: Deletion

CHANGE SUMMARY: Flame arresters or pressure-vacuum (PV) breather valves are no longer required on all protected above-ground storage tanks, only those containing Class I flammable liquids.

2015 CODE: ~~5704.2.9.7.3 Flame Arresters.~~ ~~Approved flame arresters or pressure vacuum breather valves shall be installed in normal vents.~~

CHANGE SIGNIFICANCE: The requirement for either a flame arrester or a PV breather valve on all protected above-ground storage tanks has been removed. The 2012 IFC required such a device on all protected above-ground tanks. A protected above-ground tank must pass the fire testing in UL 2085, Protected Aboveground Tanks for Flammable and Combustible Liquids. The tanks must pass the testing criteria without a flame arrester or PV breather installed.

Consider a diesel generator with an integral steel above-ground tank listed to UL 142, Steel Aboveground Tanks for Flammable and Combustible Liquids. The UL 142 tank is not required to be equipped with a flame arrester on the normal vent. However, as soon as that tank is replaced with a protected above-ground storage tank listed to UL 2085, the 2012 IFC requires a flame arrester or PV breather valve even though the tank contains the same fuel, supplies the same diesel generator, and provides a higher level of safety. This inconsistency has been corrected in the 2015 IFC.

In the 2015 IFC, there are places where a flame arrester or a PV breather valve is required, even on a protected above-ground tank. Those situations are dependent on the product contained in the tank. Specifically, the stor-

5704.2.9.7.3 continues

Flame arrester or PV breather valve required on protected above-ground tanks storing Class I flammable liquids

5704.2.9.7.3 continued   age of a Class I flammable liquid requires either a flame arrester or a PV breather valve. IFC Section 5704.2.7.3.2 applies to all types of tanks including protected above-ground storage tanks. This requirement, in pertinent part, reads as follows:

5704.2.7.3.2 Vent-Line Flame Arresters and PV Vents. Listed or approved flame arresters or PV vents that remain closed unless venting under pressure or vacuum conditions shall be installed in normal vents of tanks containing Class IB and IC liquids.

> **Exception:**   Where determined by the fire code official that the use of such devices can result in damage to the tank.

Flame arresters or PV breather valves are now required based on the contents having a higher volatility; the requirement is not based on the type of tank.

CHANGE TYPE: Addition

CHANGE SUMMARY: Requirements applicable to a hydrogen fuel gas room have been included in the IFC, providing correlation with industry standards.

2015 CODE:

SECTION 5808
HYDROGEN FUEL GAS ROOMS

5808.1 General. Where required by this code, hydrogen fuel gas rooms shall be designed and constructed in accordance with Sections 5808.1 through 5808.7 and the *International Building Code.*

5808.2 Location. Hydrogen fuel gas rooms shall not be located below grade.

5808.3 Design and Construction. Hydrogen fuel gas rooms not exceeding the maximum allowable quantity per control area in Table 5003.1.1(1) shall be separated from other areas of the building in accordance with Section 509.1 of the *International Building Code.*

5808.3.1 Pressure Control. Hydrogen fuel gas rooms shall be provided with a ventilation system designed to maintain the room at a negative pressure in relation to surrounding rooms and spaces.

5808 continues

5808
Hydrogen Fuel Gas Rooms

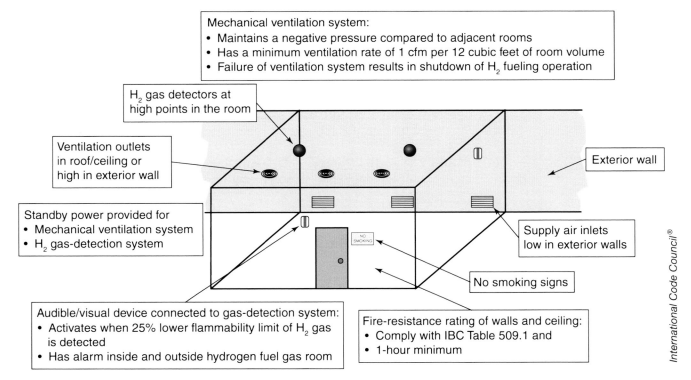

Mechanical ventilation system:
- Maintains a negative pressure compared to adjacent rooms
- Has a minimum ventilation rate of 1 cfm per 12 cubic feet of room volume
- Failure of ventilation system results in shutdown of H_2 fueling operation

H_2 gas detectors at high points in the room

Ventilation outlets in roof/ceiling or high in exterior wall

Exterior wall

Standby power provided for
- Mechanical ventilation system
- H_2 gas-detection system

Supply air inlets low in exterior walls

No smoking signs

Audible/visual device connected to gas-detection system:
- Activates when 25% lower flammability limit of H_2 gas is detected
- Has alarm inside and outside hydrogen fuel gas room

Fire-resistance rating of walls and ceiling:
- Comply with IBC Table 509.1 and
- 1-hour minimum

International Code Council®

Safety features of hydrogen fuel gas room

5808 continued

5808.3.2 Windows. Operable windows in interior walls shall not be permitted. Fixed windows shall be permitted where in accordance with Section 716 of the *International Building Code.*

5808.4 Exhaust Ventilation. Hydrogen fuel gas rooms shall be provided with mechanical exhaust ventilation in accordance with the applicable provisions of Section 2311.7.1.1.

5808.5 Gas-Detection System. Hydrogen fuel gas rooms shall be provided with an approved flammable gas detection system in accordance with Sections 5808.5.1 through 5808.5.4.

5808.5.1 System Design. The flammable gas detection system shall be listed for use with hydrogen and any other flammable gases used in the hydrogen fuel gas room. The gas detection system shall be designed to activate when the level of flammable gas exceeds 25 percent of the lower flammability limit (LFL) for the gas or mixtures present at their anticipated temperature and pressure.

5808.5.2 Gas Detection System Components. Gas detection system control units shall be listed and labeled in accordance with UL 864 or UL 2017. Gas detectors shall be listed and labeled in accordance with UL 2075 for use with the gases and vapors being detected.

5808.5.3 Operation. Activation of the gas detection system shall result in both of the following:

1. Initiation of distinct audible and visual alarm signals both inside and outside of the gas room.
2. Activation of the mechanical exhaust ventilation system.

5808.5.4 Failure of the Gas Detection System. Failure of the gas detection system shall result in activation of the mechanical exhaust ventilation system, cessation of hydrogen generation and the sounding of a trouble signal in an approved location.

5808.6 Explosion Control. Explosion control shall be provided where required by Section 911.

5808.7 Standby Power. Mechanical ventilation and gas detection systems shall be connected to a standby power system in accordance with Section 604.

**SECTION 202
GENERAL DEFINTIONS**

Gaseous Hydrogen System. An assembly of piping, devices and apparatus designed to generate, store, contain, distribute or transport a nontoxic, gaseous hydrogen-containing mixture having not less than 95-percent hydrogen gas by volume and not more than 1-percent oxygen by volume. Gaseous hydrogen systems consist of items such as *compressed gas* containers, reactors and appurtenances, including pressure regulators, pressure relief devices, manifolds, pumps, compressors and interconnecting piping and tubing and controls.

Hydrogen Fuel Gas Room. <u>A room or space that is intended exclusively to house a gaseous hydrogen system.</u>

CHANGE SIGNIFICANCE: Section 2309.3.1.2 requires hydrogen generation, compression, storage and dispensing equipment to be located in indoor rooms or areas constructed in accordance with the IBC, the *International Fuel Gas Code*, the *International Mechanical Code* and NFPA 2, *Hydrogen Technologies Code*. Based on the new definitions, this indoor room will be a hydrogen fuel gas room.

The IBC has contained criteria for the design of a hydrogen fuel gas room (previously referred to as a hydrogen cutoff room) since the 2006 edition. With the first edition of NFPA 2, *Hydrogen Technologies Code* in 2011, this change is intended to correlate the requirements in the IFC, IBC and NFPA 2.

A hydrogen fuel gas room may be classified as a Group H-2 occupancy or simply as an incidental use area within the building. The distinction will be the quantity of hydrogen gas within the room. The maximum allowable quantity in a control area (MAQ) is 1,000 cubic feet. This volume can be doubled when the building has a fire sprinkler system and doubled again when the gas is stored in gas cabinets or a gas room (see Significant Change to Table 5003.1.1(1).) Therefore, the MAQ is 4,000 cubic feet within the hydrogen fuel gas room. When the hydrogen fuel gas room qualifies as an incidental use area, IBC Table 509 requires a 1-hour separation to Group B, F, M, S and U occupancies, and a 2-hour separation to Group A, E, I and R occupancies. For rooms or control areas containing more than 4,000 cubic feet, the room would be classified as a Group H-2 occupancy and separated from the remainder of the building according to IBC Table 508.4, which could be a 2-hour, 3-hour or 4-hour requirement.

Ventilation in a hydrogen fuel gas room will conform to two criteria. First, the hydrogen fuel gas room must be maintained at a lower pressure than the surrounding portions of the building. This will assist in the containment of any gas release within the room itself. Second, an exhaust ventilation system is required that has fresh-air supply inlets at the floor level and will exhaust at a minimum rate of 1 cubic foot per minute per 12 cubic feet of room volume.

The mechanical exhaust system will either run continuously or be activated when hydrogen gas is detected by the gas detection system at a concentration above 25 percent of the LFL of hydrogen. The LFL of hydrogen is 4 percent, so detectors should activate at a hydrogen concentration of 1 percent.

When hydrogen gas is detected, an audible and visual alarm is required both inside and outside the hydrogen fuel gas room. Most likely, the outside location will be near the entry doors into the hydrogen fuel gas room. Additionally, the gas detection system will start the mechanical exhaust ventilation system in those instances where the exhaust ventilation system does not run continuously. If the gas detection system fails, the mechanical exhaust ventilation system must activate, a supervisory trouble signal must be transmitted to an approved location, and any hydrogen gas generation equipment must shut down. In order to ensure operation of the safeguards, standby power is required for both the mechanical exhaust ventilation system and the gas detection system.

Appendices A through M contain provisions for establishing boards of appeals, fire department access, fire-flow requirements for buildings, locations for fire hydrants, inspection criteria for non-compliant fire protection systems, building information signs and retroactive installation of fire sprinklers in existing high-rise buildings. They contain information concerning the classification of hazardous materials and the equivalent weights and volumes of the most commonly found cryogenic fluids. Appendices A through D and H through M do not have the force of law unless they are adopted by the jurisdiction. Appendices E through G are intended only as information sources that supplement the requirements of the *International Fire Code* (IFC) and should not be adopted. ■

APPENDIX B, B105

Fire-Flow Requirements for Buildings

APPENDIX C

Fire Hydrant Locations and Distribution

APPENDIX K

Construction Requirements for Existing Ambulatory Care Facilities

APPENDIX L

Fire Fighter Air Replenishment Systems

APPENDIX M

Retroactive Installation of Fire Sprinklers in Existing High-rise Buildings

CHANGE TYPE: Modification

CHANGE SUMMARY: Criteria have been added to Appendix B that specify the amount of reduction available for each type of fire sprinkler system and establish the method for determining the minimum water supply requirement and duration based on the reduced fire-flow requirement.

2015 CODE: B105.1 One- and Two-Family Dwellings, Group R-3 and R-4 Buildings and Townhouses. The minimum fire-flow and flow duration requirements for one- and two-family dwellings, Group R-3 and R-4 buildings and townhouses ~~having a fire-flow calculation area that does not exceed 3,600 square feet (344.5 m²) shall be 1,000 gallons per minute (3785.4 L/min) for 1 hour. Fire-flow and flow duration for dwellings having a fire-flow calculation area in excess of 3,600 square feet (344.5m²) shall not be less than that specified in Table B105.1~~ shall be as specified in Tables B105.1(1) and B105.1(2).

> **Exception:** ~~A reduction in required fire-flow of 50 percent, as approved, is allowed when the building is equipped with an approved automatic sprinkler system.~~

TABLE B105.1(1)   Required Fire-Flow for One- and Two-Family Dwellings, Group R-3 and R-4 Buildings and Townhouses

Fire-Flow Calcuation Area (square feet)	Automatic Sprinkler System (Design Standard)	Minimum Fire-Flow (gallons per minute)	Flow Duration (hours)
0 – 3,600	No automatic sprinkler system	1,000	1
3,601 – greater	No automatic sprinkler system	Value in Table B105.1(2)	Duration in Table B105.1(2) at the required fire-flow rate
0 – 3,600	Section 903.3.1.3 of the *International Fire Code* or Section P2904 of the *International Residential Code*	500	½
3,601 – greater	Section 903.3.1.3 of the *International Fire Code* or Section P2904 of the *International Residential Code*	½ value in Table B105.1(2)	1

For SI: 1 square foot = 0.0929 m², 1 gallon per minute = 3.785 L/m.

Appendix B, B105 continues

Appendix B, B105 continued

B105.2 Buildings Other Than One- and Two-Family Dwellings, Group R-3 and R-4 Buildings and Townhouses. The minimum fire-flow and flow duration for buildings other than one- and two-family dwellings, <u>Group R-3 and R-4 buildings and townhouses</u> shall be as specified in Tables ~~B105.1~~ <u>B105.2 and B105.1(2)</u>.

> **Exception:** ~~A reduction in required fire-flow of up to 75 percent, as approved, is allowed when the building is provided with an approved automatic sprinkler system installed in accordance with Section 903.3.1.1 or 903.3.1.2. The resulting fire-flow shall not be less than 1,500 gallons per minute (5678 L/min) for the prescribed duration as specified in Table B105.1.~~

B105.3 Water Supply for Buildings Equipped with an Automatic Sprinkler System. For buildings equipped with an approved automatic sprinkler system, the water supply shall be capable of providing the greater of:

1. The automatic sprinkler system *demand*, including hose stream allowance.
2. The required fire-flow.

TABLE B105.1(2)
~~**Minimum Required Fire-Flow and Flow Duration for Buildings**~~ **Reference Table for Tables B105.1(1) and B105.2**
(Portions of table not shown remain unchanged)

TABLE B105.2 Required Fire-Flow for Buildings Other Than One- and Two-Family Dwellings, Group R-3 and R-4 Buildings and Townhouses

Automatic Sprinkler System (Design Standard)	Minimum Fire-Flow (gallons per minute)	Flow Duration (hours)
No automatic sprinkler system	Value in Table B105.1(2)	Duration in Table B105.1(2)
Section 903.3.1.1 of the *International Fire Code*	25% of the value in Table B105.1(2)[a]	Duration in Table B105.1(2) at the reduced flow rate
Section 903.3.1.2 of the *International Fire Code*	25% of the value in Table B105.1(2)[b]	Duration in Table B105.1(2) at the reduced flow rate

For SI: 1 square foot = 0.0929 m², 1 gallon per minute = 3.785 L/m.
a. The reduced fire-flow shall not be less than 1,000 gallons per minute (5678 L/min).
b. The reduced fire-flow shall not be less than 1,500 gallons per minute (3785 L/min).

CHANGE SIGNIFICANCE: Two tables have been added to Appendix B. Table B105.1(1) is used to determine the required fire-flow and flow duration in Table B105.1(2) for one- and two-family dwellings, townhouses and congregate living facilities. Table B105.2 specifies the required fire-flow and flow duration in Table B105.1(2) for all other occupancies.

Group R-3 townhouses and Group R-4 buildings are now specifically included as one- and two-family dwellings with respect to determining the required fire-flow. The IRC applies to construction of one- and two-family dwellings and townhouses up to three stories in height. Under the IBC, these same facilities would be classified as Group R-3 occupancies. Additionally, the result is that all one- and two-family dwellings, townhouses, Group R-3 and Group R-4 buildings will now be treated the same when the fire code official is determining the required fire-flow, whether the buildings are built either under the IBC or the IRC.

Table B105.1(1) applies to one- and two-family dwellings, townhouses, Group R-3 and Group R-4 buildings. This table looks at two criteria:

1. Is the size of the fire-flow calculation area greater than 3600 square feet?
2. Is the building equipped with fire sprinklers?

Based on those criteria, the minimum fire-flow and flow duration can be determined. For buildings with a fire-flow calculation area exceeding 3600 square feet, reference is made to Table 105.1(2) for determination of the required fire-flow. When the building is equipped with fire sprinklers, the required fire-flow is cut in half.

Table 105.1(1) allows credit for fire sprinkler systems complying with either NFPA 13D, Standard for the Installation of Sprinklers Systems in One- and Two-family Dwellings and Manufactured Homes or Section P2904 of the *International Residential Code*. Even though it is not stated, a fire sprinkler system designed under NFPA 13, Standard for the Installation of Sprinklers Systems or NFPA 13R, Standard for the Installation of Sprinklers Systems in Low Rise Residential Occupancies should also receive credit, since these other two standards are more restrictive than NFPA 13D.

Table B105.2 applies to all buildings other than one- and two-family dwellings, townhouses, Group R-3 and Group R-4. Similar to Table 105.1(1), this table considers whether the building is equipped with fire sprinklers, but it is also concerned with what standard was used for the fire sprinkler system design. When the building does not have a fire sprinkler system, then a direct reference Table 105.1(2) is provided for determination of the fire-flow. If the fire sprinkler system was designed to NFPA 13, then a reduction to 25 percent to the tabular value in Table 105.1(2) is provided, but with a minimum fire-flow of 1,000 GPM. If the fire sprinkler system was designed to NFPA 13R, then a similar reduction to 25 percent the tabular value in Table 105.1(2) is provided, but with a minimum fire-flow of 1,500 GPM.

Table 105.2 requires 25 percent of the tabular fire-flow rate when a building is protected with a fire sprinkler system. The 2012 IFC allowed for a reduction of "up to 75 percent" of the tabular fire-flow. The end result is the same when the full 75 percent is allowed in the 2012 IFC. The fire code official can no longer limit the reduction to less than the 75 percent, since the reduction is specified and granted based on the fire sprinkler system.

Appendix B, B105 continues

Appendix B, B105 continued

Appendix B now provides specific guidance with regard to determining the minimum duration of the required fire-flow. Both Tables 105.1(1) and 105.2 state that the duration of the fire-flow is to be determined based on the calculated fire-flow after reductions are taken.

Section B105.3 specifies how to determine the water supply requirement when the fire hydrant demand and the fire sprinkler design are considered. This new section specifies that the greater of the fire sprinkler demand or the fire-flow demand will determine the minimum required water supply. When the fire sprinkler demand is considered, the hose stream allowance in NFPA 13 must be included.

Consider the following example:

- A 5-story building is proposed for construction.
- This building will be a Group B occupancy with 25,000 square feet per floor.
- The building will be Type IIA construction and equipped with a fire sprinkler system designed to NFPA 13.

Required Fire-Flow:

- Section B104.1 states that the fire-flow calculation area will be considered as the entire building since there are no fire walls separating the building into smaller fire-flow calculation areas. Therefore, the fire-flow calculation area is 125,000 square feet.
- The tabular value in Table B105.1(2) for 125,000 square feet indicates a minimum fire-flow of 5,000 GPM.
- Since the building is equipped with an NFPA 13 fire sprinkler system, Table B105.2 allows a reduction in the required fire-flow to 1,250 GPM.

Minimum Fire-Flow Requirements:

- Since the building is sprinklered, the fire sprinkler demand must be reviewed to determine the water supply requirements. For this example, assume a fire sprinkler system demand of 300 GPM (including hose stream allowance) at 50 PSI.
- The minimum fire-flow available at the fire hydrant is 1,250 GPM at 20 PSI.

Determining Minimum Water Supply:

- Section B105.3 states the requirement will be the greater of the two flow demands. Since the pressure requirements are different, a direct comparison cannot be made to determine which is greater.

- One requirement has a greater pressure, and the other has a greater volume.
- The water supply curve for the water available at the site must be able to meet both criteria: the lower volume at the higher pressure, and the higher volume at the lower pressure.

See the Water Supply Curve, which demonstrates how both flows are met.

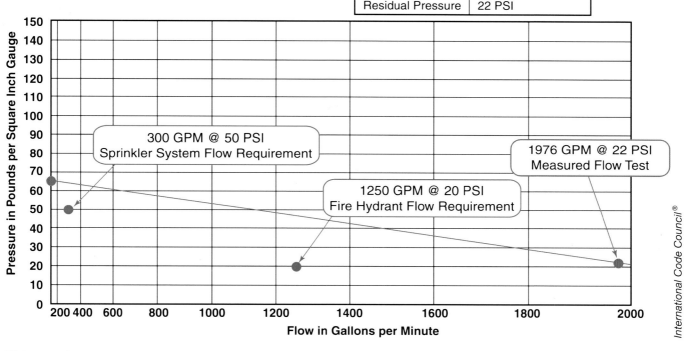

Water supply curve

Appendix C

Fire Hydrant Locations and Distribution

CHANGE TYPE: Modification

CHANGE SUMMARY: The revisions to Appendix C provide refinement of the fire hydrant spacing requirements and add footnotes that increase hydrant spacing based on the installation of an automatic sprinkler system.

2015 CODE:

APPENDIX C
FIRE HYDRANT LOCATIONS AND DISTRIBUTION

The provisions contained in this appendix are not mandatory unless specifically referenced in the adopting ordinance.

SECTION C101
GENERAL

C101.1 Scope. In addition to the requirements of Section 507.5.1 of the *International Fire Code*, fire hydrants shall be provided in accordance with this appendix for the protection of buildings, or portions of buildings, hereafter constructed or moved into the jurisdiction.

~~SECTION C102~~
~~LOCATION~~

~~**C102.1 Location.** Fire hydrants shall be provided along required fire apparatus access roads and adjacent public streets.~~

SECTION ~~C103~~ C102
NUMBER OF FIRE HYDRANTS

~~C103.1~~ C102.1 Minimum Number of Fire Hydrants for a Building ~~Available~~. The ~~minimum~~ number of fire hydrants available to a building shall not be less than ~~that listed~~ the minimum specified in Table ~~C105.1~~ C102.1.

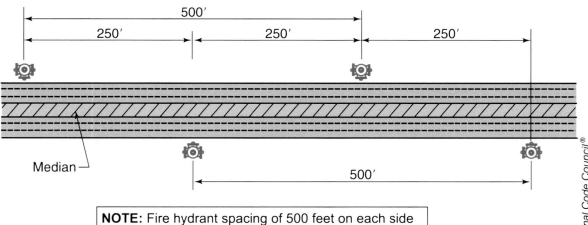

NOTE: Fire hydrant spacing of 500 feet on each side of the street is not dependent on the required fire-flow.

For SI: 1 foot = 304.8 mm.

Fire hydrant spacing along boulevards and high-volume arterial streets

International Code Council®

~~The number of fire hydrants available to a complex or subdivision shall not be less than that determined by spacing requirements listed in Table C105.1 when applied to fire apparatus access roads and perimeter public streets from which fire operations could be conducted.~~

TABLE ~~C105.1~~ C102.1 Required Number and ~~Distribution~~ Spacing of Fire Hydrants

Fire-Flow Requirement (GPM)	Minimum Number of Hydrants	Average Spacing between Hydrants [a,b,c,f,g]	Maximum Distance from Any Point on Street or Road Frontage to a Hydrant [d,f,g]
1,750 or less	1	500	250
2,000–2,250	2	450	225
2,500	3	450	225
3,000	3	400	225
3,500–4,000	4	350	210
4,500–5,000	5	300	180
5,500	6	300	180
6,000	6	250	150
6,500–7,000	7	250	150
7,500 or more	8 or more [e]	200	120

For SI: 1 foot = 304.8 mm, 1 gallon per minute = 3.785 L/m.

a. Reduce by 100 feet for dead-end streets or roads.
b. Where streets are provided with median dividers that cannot be crossed by fire fighters pulling hose lines, or where arterial streets are provided with four or more traffic lanes and have a traffic count of more than 30,000 vehicles per day, hydrant spacing shall average 500 feet on each side of the street and be arranged on an alternating basis ~~up to a fire-flow requirement of 7,000 gallons per minute and 400 feet for higher fire-flow requirements.~~
c. Where new water mains are extended along streets where hydrants are not needed for protection of structures or similar fire problems, fire hydrants shall be provided at spacing not to exceed 1,000 feet to provide for transportation hazards.
d. Reduce by 50 feet for dead-end streets or roads.
e. One hydrant for each 1,000 gallons per minute or fraction thereof.
f. A 50-percent spacing increase shall be permitted where the building is equipped throughout with an approved automatic sprinkler system in accordance with Section 903.3.1.1 of the *International Fire Code.*
g. A 25-percent spacing increase shall be permitted where the building is equipped throughout with an approved automatic sprinkler system in accordance with Sections 903.3.1.2 or 903.3.1.3 of the *International Fire Code* or Section P2904 of the *International Residential Code.*

SECTION ~~C105~~ C103
~~DISTRIBUTION OF FIRE HYDRANTS~~ FIRE HYDRANT SPACING

~~C105.1~~ C103.1 Hydrant Spacing. Fire apparatus access roads and public streets providing required access to buildings in accordance with Section 503 of the *International Fire Code* shall be provided with one or more fire hydrants, as determined by Section C102.1. Where more than one fire hydrant is required, the distance between required fire hydrants shall be in accordance with Sections C103.2 and C103.3.

Appendix C continues

Appendix C continued

C103.2 Average Spacing. The average spacing between fire hydrants shall ~~not exceed that listed in~~ be in accordance with Table ~~C105.1~~ C102.1.

> **Exception:** ~~The fire chief is authorized to accept a deficiency of up to 10 percent~~ The average spacing shall be permitted to be increased by 10 percent where existing fire hydrants provide all or a portion of the required number of fire hydrants ~~service~~.

C103.3 Maximum Spacing. ~~Regardless of the average spacing, fire hydrants shall be located such that all points on streets and access roads adjacent to a building are within the distances listed in Table C105.1 and the minimum number of hydrants are provided.~~ The maximum spacing between fire hydrants shall be in accordance with Table C102.1.

SECTION C104
CONSIDERATION OF EXISTING FIRE HYDRANTS

C104.1 Existing Fire Hydrants. Existing fire hydrants on public streets are allowed to be considered as available to meet the requirements of Sections C102 and C103. Existing fire hydrants on adjacent properties ~~shall not be considered available unless~~ are allowed to be considered as available to meet the requirements of Sections C102 and C103 provided that a fire apparatus access road extends between properties and that an easement~~s are~~ is established to prevent obstruction of such roads.

CHANGE SIGNIFICANCE: Several items have been clarified in Appendix C. Since this is an appendix, the provisions are only applicable when the appendix is specifically adopted by the jurisdiction.

Section C101.1 specifies that the requirements in Appendix C apply in addition to the requirements of Section 507.5.1. Section 507.5.1 refers to fire hydrants being located within 400 feet of all portions of a building or facility (some exceptions apply).

The exception to Section C103.2 has been revised. This revision provides a clear allowance to increase the average fire hydrant spacing by 10 percent when existing fire hydrants are considered; this spacing does not need the approval of the fire chief. This increase in spacing is justified considering that the new fire hydrants can be located wherever they are needed, while the location of the existing fire hydrants is already established.

Existing fire hydrants are further addressed in Section C104.1. This section is revised with regard to existing fire hydrants that are not located on a public way; in other words, it refers to a yard fire hydrant or on-site fire hydrant. The fire department has access to fire hydrants located along a public way. However, fire hydrants located on private property were typically installed to meet the requirements for that specific building or facility. Certainly when a fire occurs on that property, the fire department has access to those on-site fire hydrants. However, when the fire is on another property, access can be more difficult. Therefore, on-site fire hydrants can be considered only when they meet the spacing requirements and there is either a public easement or an easement for fire department use.

Table C102.1 contains three modifications to the footnotes:

1. Footnote b has been revised by deleting the 7,000 GPM threshold for arranging hydrants on an alternating basis on streets that are

difficult to cross with hose lines. This footnote now specifies that fire hydrants shall be located on both sides of the street when

- the street has a median divider that is not traversable by fire fighters pulling hose lines, or
- the street has four or more traffic lanes and a traffic count of 30,000 vehicles per day.

2. Footnote f has been added that provides an increase of 50 percent in the fire hydrant spacing when the building is protected by a fire sprinkler system installed in accordance with NFPA 13, Standard for the Installation of Sprinkler Systems.

3. Footnote g has been added that provides an increase of 25 percent in the fire hydrant spacing when the building is protected by a fire sprinkler system installed in accordance with NFPA 13D, Standard for the Installation of Sprinkler Systems in One- and Two-family Dwellings and Manufactured Homes; NFPA 13R, Standard for the Installation of Sprinkler Systems in Low Rise Residential Occupancies; or Section P2904 of the IRC.

Footnotes f and g will apply in addition to the reduction in fire-flow in Appendix B based on the installation of a fire sprinkler system.
Consider the following example:

- A 2-story building is proposed for construction.
- This building will be a Group M occupancy with 75,000 square feet per floor.
- The building will be Type IIIB construction and equipped with a fire sprinkler system.
- Two existing fire hydrants are located on the public street.

Fire-Flow Required:

- Based on Appendix B Table B105.1(2), the required fire-flow is 8,000 GPM.
- Since the building is sprinklered, the reduction in Table B105.2 is to be applied, and the required fire-flow is reduced to 2,000 GPM.

Number and Spacing of Fire Hydrants:

- Based on Appendix C Table C102.1, the maximum average spacing between fire hydrants is 450 feet, with a required fire-flow of 2,000 GPM.
- Since the building is sprinklered, Footnote f increases the fire hydrant spacing to 675 feet.
- Table C102.1 also specifies that a minimum of two fire hydrants are required.

Fire Hydrants Required:

- There are existing fire hydrants on the street that are located 450 feet apart. These two fire hydrants meet the minimum requirements in Appendix C for spacing and location.

Appendix C continues

Appendix C continued

- Section C101.1 states that the appendix requirements are in addition to Section 507.5.1. Section 507.5.1 Exception 2 requires fire hydrants within 600 feet of all portions of the exterior wall in sprinklered buildings, measured from the fire hydrant.

- In this case, additional on-site fire hydrants are required. These additional fire hydrants should be spaced no more than 675 feet from the existing fire hydrants in accordance with Table C102.1.

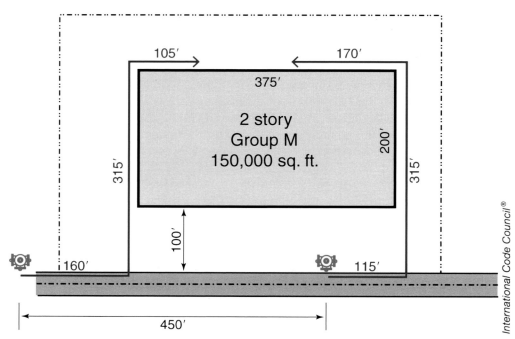

Appendix C is used with IFC Section 507.5.1 and Appendix B to determine fire hydrant locations.

International Code Council®

CHANGE TYPE: Addition

CHANGE SUMMARY: The new Appendix K addresses retroactive construction requirements for existing Ambulatory Care Facilities. The appendix requirements are in addition to the retroactive construction requirements in IFC Chapter 11.

2015 CODE:

APPENDIX K
CONSTRUCTION REQUIREMENTS FOR EXISTING
AMBULATORY CARE FACILITIES

The provisions contained in this appendix are not mandatory unless specifically referenced in the adopting ordinance.

K101.1 Scope. The provisions of this chapter shall apply to existing buildings containing ambulatory care facilities in addition to the requirements of Chapter 11 of the *International Fire Code*. Where the provisions of this chapter conflict with either the construction requirements within Chapter 11 of the *International Fire Code* or the construction requirements that applied at the time of construction, the most restrictive provision shall apply.

K101.2 Intent. The intent of this appendix is to provide a minimum degree of fire and life safety to persons occupying and existing buildings containing ambulatory care facilities where such buildings do not comply with the minimum requirements of the *International Building Code*.

Note: An outline of the remainder of Appendix K is shown here. For the entire text please refer to the 2015 IFC.

Appendix K continues

Appendix K
Construction Requirements for Existing Ambulatory Care Facilities

Existing ambulatory care facility

Appendix K continued

SECTION K102
FIRE SAFETY REQUIREMENTS
FOR EXISTING AMBULATORY CARE FACILITIES

K102.1 Separation.
K102.2 Smoke Compartments.
K102.2.1 Refuge Area.
K102.2.2 Independent Egress.
K102.3 Automatic Sprinkler Systems.
K102.4 Automatic Fire Alarm System.

SECTION K103
INCIDENTAL USES IN EXISTING
AMBULATORY CARE FACILITIES

K103.1 General.
Table K103.1 - Incidental Uses in Existing Ambulatory Care Facilities
K103.2 Occupancy Classification.
K103.3 Area Limitations.
K103.4 Separation and Protection.
K103.4.1 Separation.
K103.4.2 Protection.
K103.4.2.1 Protection Limitation.

SECTION K104
MEANS OF EGRESS REQUIREMENTS
FOR EXISTING AMBULATORY CARE FACILITIES

K104.1 Size of Doors.
K104.2 Corridor and Aisle Width.
K104.3 Existing Elevators.
K104.3.1 Elevators, Escalators, Dumbwaiters and Moving Walks.
K104.3.2 Elevator Emergency Operation.

SECTION K 105
REFERENCED STANDARDS

CHANGE SIGNIFICANCE: The appendix contains retroactive construction requirements applicable only to existing ambulatory care facilities. The IBC and IFC have only recently contained definitions, occupancy classifications and regulations specific to ambulatory care facilities. As a result, ambulatory surgery centers constructed as recently as 2010 very likely would not have been designed with defend-in-place features now required by the I-Codes. Existing facilities that were certified by the federal government as "ambulatory surgical facilities" would have the defend-in-place features included as a result of that certification. Not until the local adoption of the 2009 IBC and IFC would any defend-in-place features have been required. This appendix provides for buildings containing ambulatory care facilities to be provided with mitigation measures to allow a defend-in-place concept.

The requirements in Appendix K are designed to provide mitigation to the life hazards presented in these facilities and are not designed so that the facility will comply with current code requirements for new construction. As an example, there has been a relaxation of the requirement for a fire sprinkler system for existing facilities. Appendix K requires retroactive installation of a fire sprinkler system only when the building is of unprotected construction.

Some of the requirements found in Appendix K:

- Compliance with IFC Chapter 11 for general construction requirements in existing buildings (see Section K101.1).
- Fire barrier separation between the ambulatory care facility and the remainder of the building when there are four or more care recipients incapable of self-preservation (see Section K102.1).
- Construction of smoke compartments when the facility exceeds 10,000 square feet (see Section K102.2).
- Installation of a fire sprinkler system in Type IIB, IIIB or VB construction when four or more care recipients incapable of self-preservation, or one or more care recipients incapable of self-preservation are located on a level other than the level of exit discharge (see Section K102.3).
- Installation of a smoke detection system unless the building is protected with a fire sprinkler system (see Section K102.4).
- Protection of incidental use areas by either fire-resistance-rated separations or a fire sprinkler system (see Table K103.1).
- Provision of existing elevators with a travel distance of 25 feet or more for fire fighter operations (see Section K104.3.2).

Appendix L

Fire Fighter Air Replenishment Systems

Fill station with RIC/UAC connections located in the stairwell of an upper floor of the building

CHANGE TYPE: Addition

CHANGE SUMMARY: This new appendix provides criteria for the design, installation and testing of Fire Fighter Air Replenishment Systems (FARS) for use during firefighting operations.

2015 CODE:

APPENDIX L
REQUIREMENTS FOR FIRE FIGHTER
AIR REPLENISHMENT SYSTEMS

__The provisions contained in this appendix are not mandatory unless specifically referenced in the adopting ordinance.__

SECTION L101
GENERAL

L101.1 Scope.   Fire fighter air replenishment systems (FARS) shall be provided in accordance with this appendix. The adopting ordinance shall specify building characteristics or special hazards that establish

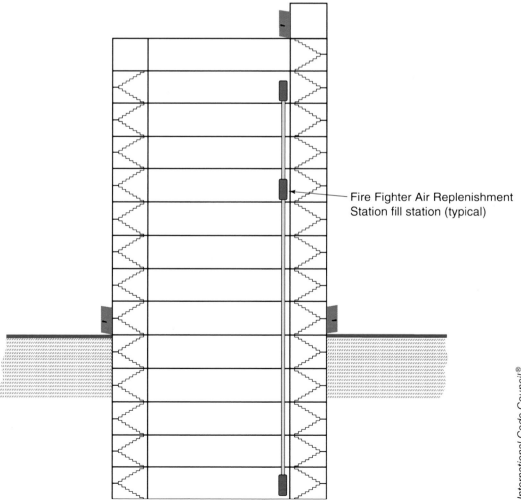

Fire Fighter Air Replenishment Station fill station (typical)

Fill stations are required on the fifth floor above and below grade and then every third floor thereafter.

International Code Council®

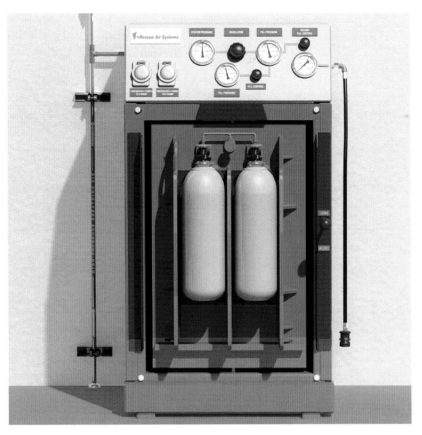

Fill station capable of simultaneously filling two cylinders within a containment system

External mobile air connection with air quality status panel

Appendix L continues

Appendix L continued   thresholds triggering a requirement for the installation of a FARS. The requirement shall be based upon the fire department's capability of replenishing fire fighter breathing air during sustained emergency operations. Considerations shall include:

1. Building characteristics, such as number of stories above or below grade plane, floor area, type of construction, and fire-resistance of the primary structural frame to allow sustained firefighting operations based on a rating of not less than 2 hours.
2. Special hazards, other than buildings, that require unique accommodations to allow the fire department to replenish fire fighter breathing air.
3. Fire department staffing level.
4. Availability of a fire department breathing air replenishment vehicle.

Note: An outline of the remainder of Appendix L is shown here. For the entire text please refer to the 2015 IFC.

SECTION L102
DEFINITIONS

L102.1 Definitions.

SECTION L103
PERMITS

L103.1 Permits.
L103.2 Construction Permit.
L103.3 Operational Permit.

SECTION L104
DESIGN AND INSTALLATION

L104.1 Design and Installation.
L104.2 Standards.
L104.2.1 Pressurized System Components.
L104.2.2 Air Quality.
L104.3 Design and Operating Pressure.
L104.4 Cylinder Refill Rate.
L104.5 Breathing Air Supply.
L104.5.1. Stored Pressure Air Supply.
L104.5.2. Retrofit of External Mobile Air Connection.
L104.6 Isolation Valves.
L104.7 Pressure Relief Valve.
L104.8 Materials and Equipment.
L104.9 Welded Connections.
L104.10 Protection of Piping.
L104.11 Compatibility.
L104.12 Security.
L104.13 Fill Stations.
L104.13.1 Location.

L104.13.2 Design.
L104.13.3 Cylinder Refill Rate.
L104.14 External Mobile Air Connection.
L104.14.1 Location.
L104.14.2 Protection from Vehicles.
L104.14.3 Clear Space Around Connections.
L104.15 Air Monitoring System.
L104.15.1 Alarm Conditions.
L104.15.2 Alarm Supervision, Monitoring and Notification.
L104.15.3 Air Quality Status Display.

SECTION L105
ACCEPTANCE TESTS

L105.1 Acceptance tests.

SECTION L106
INSPECTION, TESTING AND MAINTENANCE

L106.1 Periodic Inspection, Testing and Maintenance.

SECTION L107
REFERENCED STANDARDS

CHANGE SIGNIFICANCE: Appendix L does not require the installation of a FARS, but when such a system is installed, Appendix L provides the criteria for its design, installation, testing and maintenance.

These requirements are located in an appendix. As stated in IFC Section 101.2.1, the appendix is only enforceable when it is specifically adopted by the local jurisdiction. If the appendix is adopted, Section L101.1 states that the local adopting ordinance must provide the threshold criteria for when a FARS must be installed. The appendix provides some guidance in making the determination of which building type or design should be specified. Essentially, the concept is to provide a method for refilling self-contained breathing apparatus (SCBA) during fire fighting operations in buildings and facilities where other air delivery methods are beyond the capability of the fire department.

FARS is basically "a standpipe system for air," which provides the ability for fire fighters to refill SCBA cylinders at or near the fire-fighting location. This technology and these systems have been in use since the late 1980s.

Section L103 contains requirements for permits. A construction permit is required for the installation or modification of a FARS. Then an operational permit is required to ensure the continued maintenance and operability of the FARS.

The bulk of the appendix contains the design criteria for the FARS. Three main functions are critical to the usefulness of the FARS:

1. The availability and survivability of the system during a fire event.

2. The quality of the air supply.

3. The ability for fire fighters to refill SCBA cylinders in a reasonable length of time.

Appendix L continues

Appendix L continued

In order to provide and maintain the availability of the system, all piping and fittings must be stainless steel and designed for the operating pressures in the system. The piping must also be protected from physical damage. This could be accomplished by locating the piping within pipe chases or sleeves; however, whenever the pipe joints are concealed, the connections must be welded. All locations where connections can be made to the FARS must be secured. This would include the fill station locations as well as the external mobile air connection, if so equipped. Connections at the fill stations must be compatible with the SCBA equipment used by the fire department.

The air supply can be provided by on-site air storage, fire department mobile air supply units or a combination of both. Regardless of the type of supply, the system is to remain full and under pressure at all times. If the fire department has a mobile air unit, the FARS must include an external connection for the use by this unit. The external connection must be accessible to the mobile air unit, in a location approved by the fire chief, protected from vehicle impact, provided with adequate working space and secured. The appendix also specifies that existing FARS that are not equipped with an external mobile air connection must be retrofit with an external mobile air connection when the fire department obtains, or has access to, a mobile air supply unit.

Since the air provided in the FARS will be the breathing air for the fire fighters during a fire, it must be reliable, kept free of any contamination, and meet national standards for breathing air. Section L104.15.1 contains specific items that must be measured and then continually monitored to ensure adequate air quality. The air must not contain percentages of carbon monoxide, carbon dioxide, hydrocarbons or moisture that exceed specific thresholds. Additionally, both the oxygen and nitrogen concentrations must be within specific tolerances. To help ensure the air quality, air samples must be collected, and the samples are required to be certified by a testing laboratory prior to the FARS being placed into service. Then, samples must be collected and tested by a testing lab at least quarterly. The appendix references NFPA 1989, Standard on Breathing Air Quality for Emergency Services Respiratory Protection, which requires that the testing lab place a tag on the FARS system stating the date of the last sample testing. Additionally, a monitoring system is required to provide continual monitoring of all the air quality criteria, plus monitoring of the air pressure within the system. The air monitoring system must consist of at least two analyzers that will send a supervisory signal to an approved supervising station or to a constantly attended location. In high-rise buildings, the monitoring system must also send a signal to the fire command center. The status of the air quality monitoring must also be displayed at the external mobile air connection.

NFPA 1989 contains specific methods for determining air quality and where the test samples are to be collected. It also provides criteria for placing a FARS back into service after the air quality has been contaminated.

Accessibility of the air to the fire fighters operating within the building must be provided. The first level of air filling stations is required on the fifth floor above or below the ground floor. Additional fill stations are located at every third floor beyond the fifth floor. The fill stations are to be located in an approved location adjacent to one of the required exit stairways. The fill station must be capable of simultaneously refilling two SCBA cylinders within a 2-minute period.

Each fill station will either be equipped with an air cylinder containment system for use during the filling operation or have the ability to connect to Rapid Intervention Crew/Company Universal Air Connection (RIC/UAC) fittings on the SCBA. It is the fire code official's decision as to which method of refilling air cylinders is provided. This decision should be made after consulting with the fire chief to provide the fill method that the fire department will utilize. SCBAs complying with the 2002 edition NFPA 1981, Standard on Open-Circuit Self-Contained Breathing Apparatus (SCBA) for Emergency Services will have RIC/UAC fittings.

Several valves and components will be necessary at each fill station:

- Pressure gauge,
- Pressure-regulating device,
- Pressure-relief valve that will discharge to a safe location,
- Slow-operating valve to control the air cylinder fill rate,
- Flow-restriction device on each fill connection,
- Bleed valve to relieve pressure prior to the disconnection of the air cylinder and
- Isolation valve to shut down the air supply piping leading to the downstream fill stations.

Appendix M

Retroactive Installation of Fire Sprinklers in Existing High-rise Buildings

CHANGE TYPE: Addition

CHANGE SUMMARY: An automatic fire sprinkler system is required to be retroactively installed in existing high-rise buildings.

2015 CODE:

APPENDIX M
HIGH-RISE BUILDINGS—RETROACTIVE
AUTOMATIC SPRINKLER REQUIREMENT

The provisions contained in this appendix are not mandatory unless specifically referenced in the adopting ordinance.

SECTION M101—SCOPE

M101.1 Scope. An automatic sprinkler system shall be installed in all existing high-rise buildings in accordance with the installation requirements and compliance schedule of this appendix.

SECTION M102—WHERE REQUIRED

M102.1 High-rise Buildings. An automatic sprinkler system installed in accordance with Section 903.3.1.1 of the *International Fire Code* shall be provided throughout existing high-rise buildings.

High-rise buildings pose a significant fire risk to occupants and fire fighters.

International Code Council®

Exceptions:

1. Airport traffic control towers.

2. Open parking structures.

3. Group U Occupancies.

4. Occupancies in Group F-2.

SECTION M103—COMPLIANCE

M103.1 Compliance Schedule. Building owners shall file a compliance schedule with the fire code official no later than 365 days after the first effective date of this code. The compliance schedule shall not exceed 12 years for an automatic sprinkler system retrofit.

CHANGE SIGNIFICANCE: This new appendix requires the retroactive installation of an automatic sprinkler system in existing high-rise buildings.

Some of the more difficult structure fires to combat are those in high-rise buildings. By their very nature, high-rise fires present unique firefighting challenges that are difficult for fire fighters to mitigate. High-rise buildings create longer egress times and longer fire hose-deployment time to the seat of the fire. The very simple requirement to move fire fighters and equipment from the street level to the upper levels of the building results in complex logistics and the need for additional relief personnel.

A prime example of the hazard present in high-rise buildings is the One Meridian Plaza fire in Philadelphia on February 23, 1991. The fire occurred on the 22nd floor of the 38-story Meridian Bank Building and burned for more than 19 hours. The fire resulted in three fire fighter fatalities and injuries to 24 fire fighters. The 12-alarm fire brought 51 engine companies, 15 ladder companies, 11 specialized units and over 300 fire fighters to the scene. It was the largest high-rise office building fire in modern American history, completely consuming eight floors of the building, and was only controlled when it reached a floor that was protected by automatic fire sprinklers.

Today's requirements in the IBC and IFC mandate complete fire sprinkler protection and a variety of other safety features in new high-rise buildings. Many older high-rise buildings lack automatic sprinkler systems and other fire protection features necessary to protect the occupants, emergency responders and the structure itself. This appendix does not require full compliance with the current codes, but of all the possible fire protection features available, a fire sprinkler system provides the greatest benefit to enhance the level of safety.

This requirement is located in an appendix. According to Section 101.2.1, it is applicable only if the appendix is specifically included in the local regulations adopting the IFC. There are four types of buildings that are exempted from the requirement. Airport traffic control towers, open parking structures, and Groups F and U occupancies have traditionally not been required to install fire sprinklers. Although revisions to Chapter 4 of the 2015 IBC will now require all new airport traffic control towers to be sprinklered, this appendix will allow existing unsprinklered airport traffic control towers to continue in operation without a fire sprinkler system.

Appendix M continues

Appendix M continued

The local application of this requirement will need to follow a procedure such as

1. Appendix M must be included and adopted by the jurisdiction;
2. The fire code official must approach the owner/manager of the unsprinklered high-rise building and identify the need for the building to be retrofitted with a fire sprinkler system; and
3. A time schedule must be established for the installation of the fire sprinkler system, which cannot exceed 12 years.

INDEX

 **INTERNATIONAL CODE COUNCIL®**

People Helping People Build a Safer World®

Dedicated to the Support of Building Safety Professionals

An Overview of the International Code Council

The International Code Council is a member-focused association. It is dedicated to developing model codes and standards used in the design, build and compliance process to construct safe, sustainable, affordable and resilient structures. Most U.S. communities and many global markets choose the International Codes.

Services of the ICC

The organizations that comprise the International Code Council offer unmatched technical, educational and informational products and services in support of the International Codes, with more than 200 highly qualified staff members at 16 offices throughout the United States. Some of the products and services readily available to code users include:

- **CODE APPLICATION ASSISTANCE**
- **EDUCATIONAL PROGRAMS**
- **CERTIFICATION PROGRAMS**
- **TECHNICAL HANDBOOKS AND WORKBOOKS**
- **PLAN REVIEW SERVICES**
- **ELECTRONIC PRODUCTS**
- **MONTHLY ONLINE MAGAZINES AND NEWSLETTERS**

- **PUBLICATION OF PROPOSED CODE CHANGES**
- **TRAINING AND INFORMATIONAL VIDEOS**
- **BUILDING DEPARTMENT ACCREDITATION PROGRAMS**
- **EVALUATION SERVICE FOR CODE COMPLIANCE**
- **EVALUATIONS UNDER GREEN CODES, STANDARDS AND RATING SYSTEMS**

Additional Support for Professionals and Industry:

ICC EVALUATION SERVICE (ICC-ES)

ICC-ES is the industry leader in performing technical evaluations for code compliance, providing regulators and construction professionals with clear evidence that products comply with codes and standards.

INTERNATIONAL ACCREDITATION SERVICE (IAS)

IAS accredits testing and calibration laboratories, inspection agencies, building departments, fabricator inspection programs and IBC special inspection agencies.

NEED MORE INFORMATION? CONTACT ICC TODAY!
1-888-ICC-SAFE | (422-7233) | www.iccsafe.org

14-09141

INTERNATIONAL CODE COUNCIL

People Helping People Build a Safer World®

Growing your career is what ICC Membership is all about

As your building career grows, so does the need to expand your code knowledge and job skills. Whether you're seeking a higher level of certification or professional quality training, Membership in ICC offers the best in I-Code resources and training for growing your building career today and for the future.

- **Learn** new job skills to prepare for a higher level of responsibility within your organization
- **Improve** your code knowledge to keep pace with the latest International Codes (I-Codes)
- **Achieve** additional ICC Certifications to open the door to better job opportunities

Plus, an affordable ICC Membership provides exclusive Member-only benefits including:

- Free code opinions from I-Code experts
- Free I-Code book(s) to new Members*
- Access to employment opportunities in the ICC Career Center
- Discounts on professional training & Certification renewals
- Savings of up to 25% off on code books & training materials
- Free benefits - Governmental Members: Your staff can receive free ICC benefits too*
- And much more!

Join the International Code Council (ICC) and start growing your building career now! Visit our Member page at www.iccsafe.org/membership for an application.

*Some restrictions apply. Speak with an ICC Member Services Representative for details.

14-09142

ICC EVALUATION SERVICE

APPROVE WITH
CONFIDENCE

- ICC-ES® Evaluation Reports are the most widely accepted and trusted in the nation.

- ICC-ES is dedicated to the highest levels of customer service, quality and technical excellence.

- ICC-ES is a subsidiary of ICC®, the publisher of the IBC®, IRC®, IPC® and other I-Codes®.

We do thorough evaluations. You approve with confidence.

Look for the ICC-ES marks of conformity before approving for installation

www.icc-es.org | 800-423-6587

14-09094

Subsidiary of ICC

INTERNATIONAL CODE COUNCIL®

GET IMMEDIATE DOWNLOADS OF THE STANDARDS YOU NEED

Browse hundreds of industry standards adopted by reference. Available to you 24/7!

Count on ICC for standards from a variety of publishers, including:

ACI	CPSC	GYPSUM
AISC	CSA	HUD
ANSI	DOC	ICC
APA	DOJ	ISO
APSP	DOL	NSF
ASHRAE	DOTn	SMACNA
ASTM	FEMA	USC
AWC	GBI	

DOWNLOAD YOUR STANDARDS TODAY!
SHOP.ICCSAFE.ORG